Biosensors: Recent advances and mathematical challenges

Editors:

Johann F. Osma
Margarita Stoytcheva

Editors:

Johann F. Osma, University of Los Andes, Colombia

Margarita Stoytcheva, Autonomous University of Baja California, Mexico

jf.osma43@uniandes.edu.co, margarita.stoytcheva@uabc.edu.mx

ISBN: 978-84-941872-0-9

DL: B-6486-2014

DOI: http://dx.doi.org/10.3926/oms.98

© OmniaScience (Omnia Publisher SL) 2014

Cover design: OmniaScience

Cover photo: © Belkin & Co - Fotolia.com

Index

Section 2. Mathematical methods for biosensors data analysis and response modeling

Section 3. Disturbances modeling in biomedical sensors systems

Prologue

Many researchers and research groups around the globe are dealing with interdisciplinary problems and multidisciplinary groups; but especially those that deal everyday with biosensors have the feeling of not having enough personnel with the sufficient background to cover all topics. With this feeling in our minds, we decided to invite wonderful researchers from different parts to design this book, not as a conventional biosensor's book, but as an interesting journey in the complex world of biosensors. A close look to some recent and key advances on the topic, but at the same time tips through novel mathematical modelling to improve our work, and last but not least, tricks to fool those undesired electrical disturbances that commonly messes our daily work can be found in these pages.

Feel free to navigate the book through its chapters, once in your hands you will feel eager to learn more about biosensors. It has been an experience to put all this together; but now, with the book ready, we can say that this trip has being a wonderful adventure. That is why want to thank all the authors for their great contributions, the Editorial and Irene for their constant input, and to you, the reader, for being part of this book.

Johann F. Osma

Margarita Stoytcheva

Introduction

The progress of the biotechnology and of the material science, associated with the modern principles of transduction of the chemical information initiated the development, during the 60th of the XX century, of new analytica devices called biosensors.

According to the IUPAC definition, "a biosensor is an integrated receptor-transducer device, which is capable of providing selective quantitative or semi-quantitative analytical information using biological recognition element".

Although of the large expansion of the biosensors dedicated research, some challenges still remain to fully exploit the biosensors potential. Therefore, this book is intended to provide an overview of the current state of the art and of the emerging trends in the field of the biosensors. It includes 10 chapters, organized in three sections.

The first book section "Recent advances in biosensors development and application" covers issues associated with the use of selected inorganic nanomaterials and their composites in biosensors (Chapter 1), the monitoring of the bacterial membrane formation and its surface characterization applying QCMD and AFM techniques (Chapter 2), and the thermodynamics of the whole cells-substrate interaction (Chapter 3). Chapters 4 and 5 deal with some specific applications of the electrochemical biosensors, namely for cyanotoxins and phenolics pollutants quantification. The electrochemical dopamine determination is the subject of Chapter 6. Nevertheless, in this case, a chemically modified electrode has been used.

The second book section "Mathematical methods for biosensors data analysis and response modeling" provides information related to the application of artificial neural network as a multivariate calibration tool for multianalyte determination (Chapter 7), and the application of machine learning methods for modeling the response of a glucose oxidase and of an acetylcholinesterase based sensors (Chapters 8 and 9).

The third book section addresses disturbances modeling in biomedical sensors systems, applying various methodologies to simulate the distortions and strategies to mitigate their influence.

The book provides significant and up-to-date information on the diverse aspects of the biosensors related researches. The multi-faceted approach and the multi-authored character of the edition enrich it and make it compelling for a large range of specialists.

Johann F. Osma

Margarita Stoytcheva

Section 1

Recent Advances in Biosensors
Research and Application

Chapter 1

Use of functional metalic nanostructures in biosensors

Rukan Genc

Mersin University, Chemical Engineering Department, Turkey.

rukangnc@gmail.com

Doi: http://dx.doi.org/10.3926/oms.138

Referencing this chapter

Genc, R. (2014). Use of functional metalic nanostructures in biosensors. In M. Stoytcheva & J.F. Osma (Eds.). *Biosensors: Recent Advances and Mathematical Challenges*. Barcelona, España: OmniaScience, pp. 15-40.

1. Introduction

Nanotechnology is the scientific area which includes materials with at least one dimension smaller than 100 nm. Not only the size but also the functionality at nano or macroscale is essential for a material to be considered as nanomaterial. Beyond the vastly increasing application areas, nanostructures are becoming superstars of analytical chemistry centered research, which are devoted to nanotechnology in order to achieve higher sensitivity, lower the price and detection limits (Jianrong, Yuqing, Nongyue, Xiaohua & Sijiao, 2004; Merkoçi, 2013). The integration of nanomaterials into biosensing systems represents one of the hottest topics of the todays research. Science direct search for the keywords "Nanomaterial AND Biosensor" came up with 1380 article and 138 reviews in Chemistry, Analytical Chemistry and Material Science Journals in the last five years, with a logarithmic increase in report number per year. This increased demand is due to unique properties possessed by nanoscale materials as a result of their tailorable nanosize and structure which offer excellent prospects for designing novel sensing systems and enhancing the performance of the biorecognition element with proved electronic signal transduction (Jianrong et al., 2004; Merkoçi, 2009).

The engagement of nanomaterials with sensing devices and/or in building sensor platforms generates novel interfaces that enable optical or electrochemical detection of the biomolecule of whim with enhanced sensitivity (Jianrong et al., 2004; Michalet, Pinaud, Bentolila, Tsay, Doose, Li et al., 2005; Putzbach & Ronkainen, 2013). Nanomaterial can enrolled as either effective optical (Byun, 2010) (Staleva, Skrabalak, Carey, Kosel, Xia, Hartland et al., 2009), fluorescent (Michalet et al., 2005), and catalytic labels (Saha, Agasti, Kim, Li & Rotello, 2012) (Costa, De la Escosura-Muñiz & Merkoçi, 2010) with signal amplifying features to increase the sensor sensitivity. In the meantime, they could be functional building blocks for functional, highly catalytic and conductive sensor platforms (Zhang, Carr & Alocilja, 2009).

The number of different type of nanostructures is increasing and wide range of nanoscale materials of different sizes, shapes and compositions are now available (Burda, Chen, Narayanan & El-Sayed, 2005; Kim, Hiraiwa, Lee & Lee, 2013; Xia & Lim, 2010). From those, mainly, nanomaterials can be divided into three main classes depending on the material they are made up of: i) inorganic nanoparticles where the core material is an inorganic element or mixture (e.g; gold, silver, TiO_2, ZnO, CdS and so on) (Lee, Sung & Park, 2011), ii) organic soft nanomaterials which are formed of organic materials including lipids, peptides, genetic material, (Genç, Ortiz & O'Sullivan, 2009; Hartgerink, Beniash & Stupp, 2001) and finally, iii) nanocomposites which are based on both organic and inorganic materials, for example, magnetosomes (Goldhawk, Rohani, Sengupta, Gelman & Prato, 2012), metal coated carbon nanotubes (Jiang, Zhang, Wang, Xu & Li, 2011) and peptide amphiphiles (Templates, 2002). However, this chapter will only cover the state of the art associated to the advantages offered by different types of inorganic nanomaterials and their composites. Collection of literature on challenges and drawbacks, and real world applications of these kinds of nanomaterials in biosensor development, including current status and future prospects will also be served to the readers' interest.

2. Inorganic Nanostructures

2.1. Gold and Silver nanoparticles

Gold and silver are the most widely employed metals in nanotechnology, which their use goes back to ancient times (e.g. for staining ruby glass (Himmelhaus & Takei, 2000), potion making, as healer (Vohs & Fahlman, 2007), bactericide (Matsumura, Yoshikata, Kunisaki & Tsuchido, 2003) and medicine (Mahdihassan, 1984)). Today, scientific authorities have also appreciated gold and silver due to their size and shape depended unique optical, electrochemical and electrical properties which can be altered towards the needs of specific applications including targeted drug therapy (Wang, Sui & Fan, 2010) and design of diagnosis and screening tools (Esseghaier, Ng & Zourob, 2012). In particular, these nanoparticles have great potential as contrast agents, fluorescent labels, and functional sensing platforms for the optical imaging and biosensing (Guo & Wang, 2007; Merkoçi, 2009, 2013).

The nanosized gold and silver show a localized surface plasmon resonance (SPR) property, which makes them perfect tools for optical sensing (D'Agata & Spoto, 2013). This size and shape depended physical plasmonic phenomenon can be explained as the absorption of the light in the visible spectrum due to the photoexcitation driven oscillation of conduction electrons inside the metal nanostructure (Mody, Siwale, Singh & Mody, 2010). A complete theoretical insight to the phenomenon was given in the book edited by Merkoçi (Merkoçi, 2013) and some other comprehensible reviews (Sato, Hosokawa & Maeda, 2007; Guerrini & Graham, 2012; Saha et al., 2012; Staleva et al., 2009; West & Halas, 2003).

The strong change in the absorption spectrum of the colloidal solution of metal nanostructures with varied sizes and shapes can be distinguishable even by naked eye. Beside these, the location of the SPR band is strongly sensitive to the environment of the nanostructure (Burda et al., 2005). A small change on the surface chemistry as a result of interaction with the target molecule will result in a strong absorbance shift from red to blue, which can easily be detected by optical methods. There are several attempts in the literature reporting optical detection of disease markers in which the detection sensitivity was altered by changing the morphology and the size of metal nanostructures, or combining with other strategies (Byun, 2010; Kim, Yoo, Park, Yoshikawa, Tamiya, Park et al., 2011; West & Halas, 2003). However, we will limit the discussion to recent examples of such achievements.

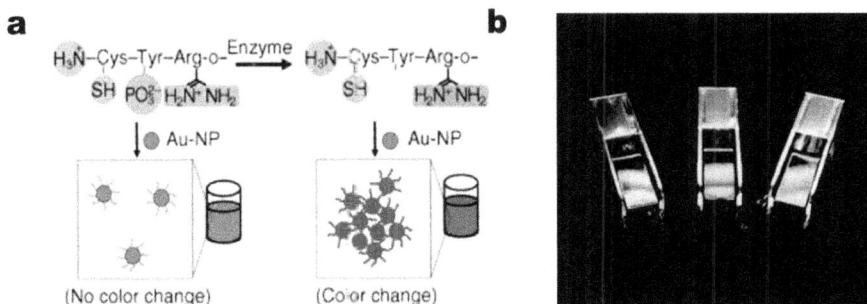

Figure 1. Gold nanoparticle based optical biosensors. a) Solution based Alkaline phosphatase (ALP) assay depicting color changes during peptide-induced Au-NP aggregation. b) Optical solid surface biosensor: cap shaped gold nanoparticles are deposited on polystyrene surface. Images are modified from Choi, Ho and Tung, 2007; Himmelhaus and Takei, 2000.

The most common strategy is the use of colloidal solution of nanoparticles which are previously bioconjugated with the capture molecule (single strand DNA, antibody/antigen, receptor, peptide and aptamer vs.) (Sato et al., 2007; Saha et al., 2012; Zhang, Guo & Cui, 2009). Interaction of the analyte with the particle surface will eventually lead a shift in SPR band which later be correlated with the analyte concentration (Figure 1a). However, the smallest working volume is limited to as much as hundreds of microns and the sensitivity of this bulk system is still poor. Same approach can also be conducted to the surfaces, where nanoparticle embedded glass surfaces can be used (Figure 1b) as a biosensor platform (Himmelhaus & Takei, 2000). This strategy is more suited for Lab on a Chip applications since it provides the opportunity to produce multi-compartmental sensors with the ability to screen multivariate analysis with minimized sample volume (Kim et al., 2011). The limiting factors of working with such system are, first, obtaining the uniform metal nanoparticle deposited surface, and later surface biofunctionalization with the capture molecule and blocking agents, since most of the absorbance shifts may occur in these very first steps depending on the size of the capture molecule, a it effects the measurement sensitivity and narrows the linear range of sample concentration.

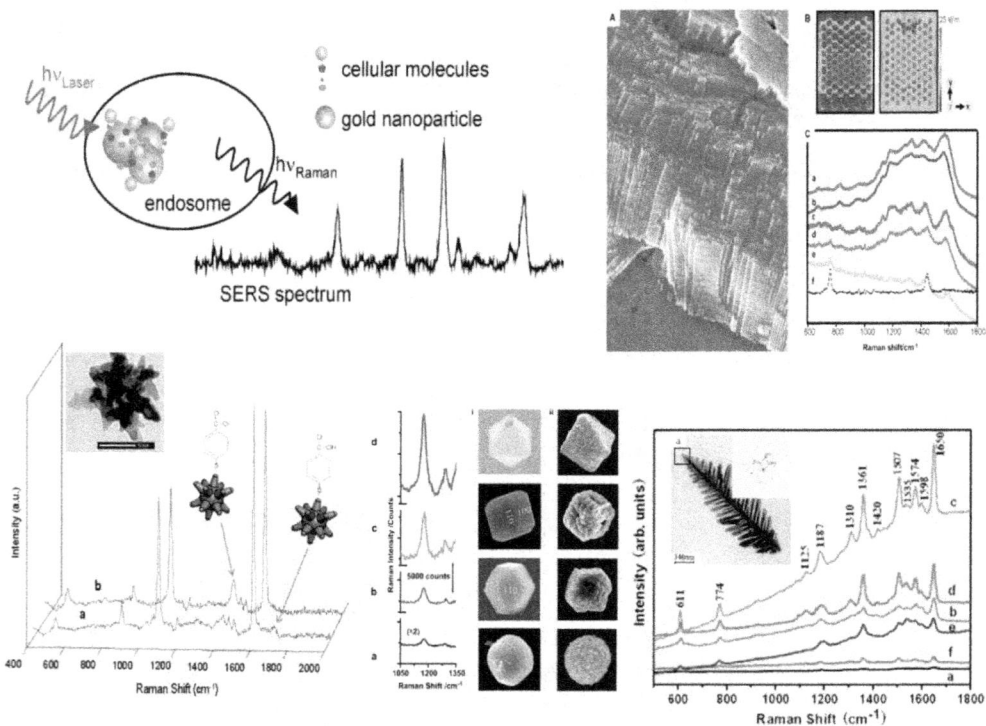

Figure 2. Introduction of different shaped gold and silver nanostructures to LSPR biosensors. Image reformatted from Alvarez-Puebla, Zubarev, Kotov and Liz-Marzán, 2012; Fang, Liu and Li, 2011; Khoury and Vo-dinh, 2008; Kneipp, Kneipp, McLaughlin, Brown and Kneipp, 2006; Yi, Chen, Chen, Luo, Wu, Yi et al., 2012.

Another technology that uses the localized SPR of nanoparticles for sensing is the surface enhanced raman scattering (SERS) (Tripp, Dluhy & Zhao, 2008) which is a powerful analytical tool provides quantification of trace constituents in biological and environmental samples, in many cases, with single-molecule sensitivity (Gunawidjaja, Kharlampieva, Choi & Tsukruk, 2009). Use of suitable nano-patterned surfaces or functionalized surfaces made up of metal nanoparticles can improve the raman signal up to 10^{15} fold (Pustovit & Shahbazyan, 2006). By this mean, both gold and silver are considered as noble materials for nanoparticle based SERS, however, they have drawbacks, too. For example, while silver shows larger SERS enhancements at visible and near IR region, gold possesses several advantageous properties over silver, such as higher performance, inertness, facile preparation and surface modification (Pustovit & Shahbazyan, 2006).

SERS platforms can be built from either deposited nanoparticles or graphed using nanolithography. Nanoparticles with tunable gaps by use of various linkers and bridge molecules alters the SERS signal more effectively while sensor surfaces constructed with these strategies still shows lack of reproducibility. Thus, there are several attempts for constructing sensor platforms with enhanced signal, reproducibility, sensitivity and stability via conducting different shaped nanostructures, as depicted in Figure 2 (nanocages, nanorod, nanowires, nanostars and dentric nanostructures) (Alvarez-Puebla et al., 2012; Fang et al., 2011; Kattumenu, Lee, Tian, McConney & Singamaneni, 2011; Khoury & Vo-Dinh, 2008; Ranjan & Facsko, 2012; Schütz, Steinigeweg, Salehi, Kömpe & Schlücker, 2011; Yi et al., 2012), as well as, developing new generation composites and hybrids with other structures including carbon nanotubes (Beqa, Singh, Fan, Senapati & Ray, 2011; Jiang et al., 2011), silica (Guerrini & Graham, 2012; Gunawidjaja et al., 2009) and quantum dots (Rumyantseva, Kostcheev, Adam, Gaponenko, Vaschenko, Kulakovich et al., 2013).

Besides the optical properties of the gold and silver nanoparticles, they are also appreciated members of electrochemical sensors. As defined by IUPAC, electrochemical sensors are "Self-contained integrated devices, which are capable of providing specific quantitative or semi-quantitative information using a biological recognition element (biochemical receptor) which is retained in direct spatial contact with an electrochemical transduction element." (Thévenot, Toth, Durst & Wilson, 1999) and are still lead the way for biomolecule detection among the other analytical approaches.

One of the known problems of electrochemical sensing of proteins is to find the suitable mediator for building the right connection between the active side of the protein and the electrode. Deposited gold nanoparticles on sensor surfaces were reported as they act like mediator which comforts the protein's free orientation and leads direct electron transfer through quantum tunneling effect. (Guo & Wang, 2007). Not only their conductivity but also catalytically properties due to high surface area, and electron density are abundantly employed to sensor technology. When colloidal gold at acidic conditions oxidized electrochemically, it produces reduced $AuCl_4^-$ ions, and in this way analyte can be quantified by measuring the change in the reduction current (Omidfar, Khorsand & Azizi, 2013). Reports consist of several other novel strategies using various types of gold nanostructures for the detection of molecules such as glycoproteins, growth factors, viruses and even metabolites directly from the cells, are exist in the literature and reader is directed to very valuable recent reviews and articles on the progress so far. (Ankri, Peretz, Motiei, Popovtzer & Fixler, 2012; Cheng, Huan & Yu, 2012; De La Escosura-Muñiz & Merkoçi, 2011; Liu, Wang & Lin, 2006). In some of these strategies, systems combining gold nanoparticles with other functional nanomaterials (carbon nanotubes, denritic nanoparticles, silica nanoparticles and magnetic nanoparticles, so on) were reported in order to increase the sensitivity and analyte-ligand interaction, as well as maintaining the physiological environment. A very interesting example is the study of Costa et al. in which they used gold nanoparticles as label and redox-element in magnetoimmunoassays for Humana IgG detection (Costa et al., 2010). While magnetic nanoparticles used to pre-concentrate the analyte and deposit them on sensor surface, gold nanoparticle label provided catalytic system revealing Hydrogen Evolution Reaction (HER) in acidic conditions (Figure 3).

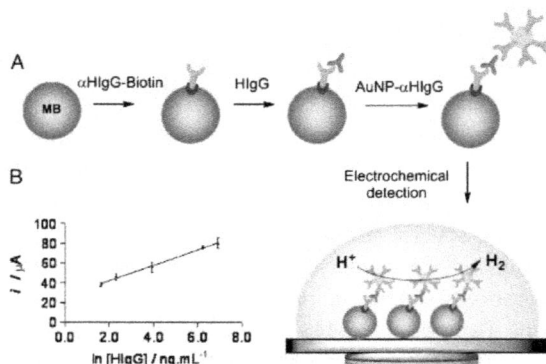

Figure 3. Gold nanoparticles (AuNPs) catalyzed Hydrogen Evolution Reaction (HER) at acidic environment which is for electrochemical detection of human IgG (Costa et al., 2010).

Silver nanoparticles and other nanostructures including nanowires and dendritic forms on the other hand, displays advantageous features same as gold, including easy dissolution, oxidization without additional chemicals and better electrochemical properties providing signal amplification (Yang, Hua, Chen & Tsai, 2013; Yi et al., 2012). For both gold and silver based nanostructures, homogenous deposition of them to the electrode surface and effectiveness of the functionalization are critical points to sustain sensor selectivity and the performance.

2.2. Quantum Dots

Fluorescent labels, mostly organic dyes such as rhodamine derivatives, fluorescein and cyanine dyes, are indisputable members of biosensors and bioassays. However, due to limitations of traditional fluorophores, for example; short band range, low stability by means of photo-bleaching and photo-oxidation, and short fluorescence lifetime, attentions are directed to a new generation inorganic, semi-conductive florescent nanoparticle, so called quantum dots (QDots) (Resch-Genger, Grabolle, Cavaliere-Jaricot, Nitschke & Nann, 2008). These nanoparticles show unique fluorescence properties which overcomes aforementioned limitations of organic fluorophores. Compared to the traditional dyes, inorganic QDots have higher resistance to the photobleaching, less sensitive to their local environment with narrower and symmetric fluorescence spectra which offer longer times of measurement (Bang & Kamat, 2009; Michalet et al., 2005; Resch-Genger et al., 2008). Most importantly, various types of quantum dots which emit light with exceptionally pure and bright colors with changing bulk band gap energies can easily be obtained by simply tuning their size or the formulation (Figure 4) (Asokan, Krueger, Alkhawaldeh, Carreon, Mu, Colvin et al., 2005). Quantum dots synthesized from several semiconductor materials in which they can be classified as II-VI type QDots from CdS, CdSe,

CdTel; III-V type QDots from InP, InAsl and IV-VI type QDots from: PbSel (Michalet et al., 2005). Besides, with their core-shell analogues, they are widely used as label in bioimaging, diagnosis, and drug delivery/tracking (Asokan et al., 2005; Dabbousi, Rodriguez-Viejo, Mikulec, Heine, Mattoussi, Ober et al., 1997). Combining QDots as label for different targets offers quite versatile tool for designing multiplex detection systems (Koehne, Chen, Cassell, Ye, Han, Meyyappan et al., 2004; Lowe, Dick, Cohen & Stevens, 2011; Rissin, Kan, Song, Rivnak, Fishburn, Shao et al., 2013). Introduction of nanomaterials to barcode amplification detection methods enables multivariant detection of proteins at submicron concentrations (Hill & Mirkin, 2006; Nam, Wise & Groves, 2005; Zhang, Carr & Alocilja, 2009). QDots with wide range of colors are versatile tools for barcoding. For example, a barcode assay system combining advantages of QDots and magnetic beads together in a single microfluidic system was reported by Gao et al, where they successfully demonstrated the multiplex recognition of four genetic targets; HIV, hepatitis B (HBV) and syphilis (Treponema pallidum) with a detection limit reaching to nM (Gao, Lam & Chan, 2013).

Another important strategy where QDots are effectively used is Förster resonance energy transfer (FRET) based sensing systems. The FRET phenomenon emerges as a result of a distance-dependent interaction which chromophore (a donor and an acceptor molecule) at close proximity (Wegner, Lanh, Jenrings, Jain, Fairclough et al., 2013). In these systems, there is a signal ON-OFF switch which regulated by changes in the close proximity (maximum 10 nm) between FRET pairs as a result of target-host interaction (Lowe et al., 2011). In this scenario, while the quantum dots are mostly act as donor; acceptor can be either QDots, fluorescent dyes, proteins or metal and metal nanoparticles. A multiplex analysis time gated FRET based detection system for measuring protease activity was reported by use of single QDot as both donor and acceptor in contrary to the traditional multiplex analysis devices with multiple QDot label (Algar, Malanoski, Susumu, Stewart, Hildebrandt & Medintz, 2012). Qdot were coated with peptidic substrates designed for trypsin and chymotrypsin which labeled with florescent Terbium, Tb, (donor) and another fluorophore A_{647} (acceptor). As a result, real time protease activity was measured through the cleavage of peptide substrates resulting in photoluminescence intensity change in Tb, QDot, and A647 giving two distinct analytical signals (Figure 4a). The designed approach offers a versatile real time detection strategy with full-package of information on the protease activity including substrate/product concentrations, initial reaction rates, enzyme-substrate specificity constants (Kc/Km), and apparent inhibition constants with no significant cross contamination (Ki). This type of FRET based strategies can be extended to several other examples (Merkoçi, 2009).

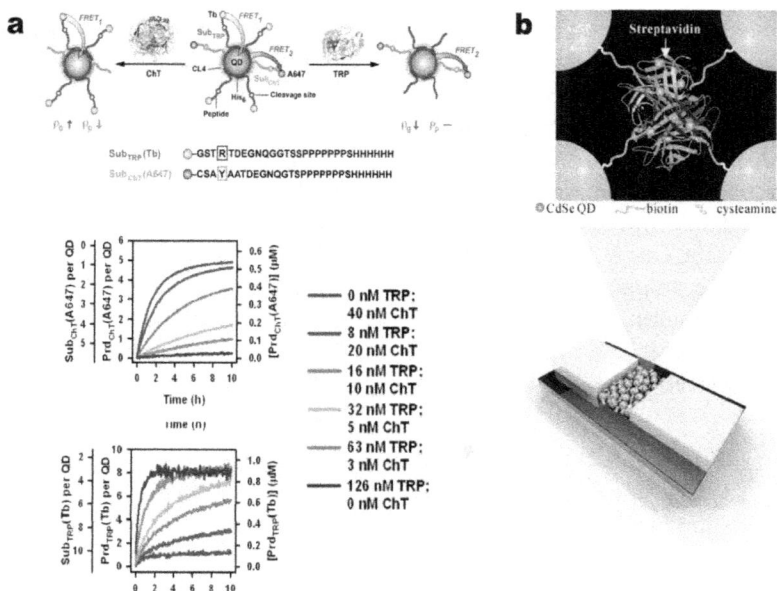

Figure 4. a) (Upper) Schematic of a time-gated FRET for multiplexed protease sensing using single QDot and its surface functionalisation with two labeles with either Tb or A647, serve as substrates for different proteases, where the cleavage sites are highlighted in the peptide sequences. (Buttom) Fluorescence signal at mixtures of different concentrations of TRP (8–126 nM) and ChT (3–40 nM) (Algar et al., 2012) b) reformed image of CdSe Quantum Dots enhanced nanogap devices for analysis of streptavidine-biotin pair as model target system (Yu, Chen, Wei, Liu, Yu & Huang, 2012).

Last but not the least, as explained in Section 2.1, in quantum gap regulated biosensor studies, the nanogaps in between metal nanoparticles are used as conducting bridges allowing direct flow of electrons (Li, Hu & Zhu, 2010; Lin, Bai & Huang, 2009). Once the analyte paired with its substrate/ligand in the nanogap (Figure 4b), it quickly reforms a strong electrical signal, which is applicable to the detection of any target from atoms to macromolecules (Li et al., 2010). Regarding to in situ growth of nanoparticles on the electrode surface, the aim is obtaining nanoparticles localized in close proximity enough for the electron tunneling. However, the dilemma of this methodology is that, to obtain nanogaps close enough (<5 nm), high number of nanoparticle is required and this naturally drives nanoparticles to agglomeration which reveals heterogeneous film on the electrode (Zhu, Zhang, Guo, Wang, Liu & Zhang, 2012). Whereas, introducing quantum dots to the nanogap based systems would overcome this drawback by propagating a supporting bridge conducting the nanoparticles grown at considerably lower concentrations, and thus, localized with larger gap interval (Yu et al., 2012). By this way, QDots could not only build a connecting bridge between nanoparticles and enrolled in signal enhancement, but also, they can regulate the electron tunneling path first between nanoelectrodes (AuNPs) and in big picture between the microelectrodes (Dahnovsky, Krevchik, Semenov, Yamamoto, Zhukovsky, Aringazin et al., 2005).

Despite many advantages of QDots as fluorescent label and signal transducer and/or enhancer in biosensing/bioimaging approaches, there are some negative aspects to be clarified before widening their use in clinical applications. They might be cytotoxic leading cell apoptosis, they are not biodegradable, and accumulation of them in certain tissues and organs might result in further problems (Amna, Van Ba, Vaseem, Hassan, Khil, Hahn et al., 2013; Knight & Serrano, 2006; Winnik & Maysinger, 2013). Moreover, their optical properties can randomly vary over time. A phenomenon regarding to these variations is quantum-dot "blinking" which is associated with mobile charges on the nanoparticle surface (Frantsuzov & Marcus, 2005; Lee & Osborne, 2009). However, indispensable Pros of quantum dots over organic dyes shadow any of these negative aspects and carry them among the top nanomaterials list for future developments in biosensor development.

2.3. Magnetic Nanoparticles

Magnetic nanoparticles (MNP) which show size depended superparamagnetism have possessed great success in separation processes. These nanoparticles with no magnetic memory -they exhibit their magnetic behavior only when an external magnetic field is applied- have the advantage over bulk counterparts with no agglomeration after a proper surface modification or coating (Mody et al., 2010). Polymers, lipids, silica and amphipathic molecules have been successfully used as coating material for not only prevent from precipitation but also favor the further functionalization of them with active molecules (Alwi, Telenkov, Mandelis, Leshuk, Gu, Oladepo et al., 2012; Floris, Ardu, Musinu, Piccaluga, Fadda, Sinico et al., 2011; Goldhawk et al., 2012; Kim, Mikhaylova, Wang, Kehr, Bjelke, Zhang et al., 2003; Mikhaylova, Kim, Bobrysheva, Osmolowsky, Semenov, Tsakalakos et al., 2004; Sawant, Sawant, Gultepe, Nagesha, Papahadjopoulos-Sternberg, Sridhar et al., 2009). Among all types of magnetic nanoparticles, iron oxides (magnetite and maghemites) are found to be more potent for analytical uses due to their being inert and also their FDA certified biocompatibility, and low level of toxicity (Gupta & Gupta, 2005; Yu, Jeong, Park, Park, Kim, Min et al., 2008). Although, most of the studies on the use of magnetic nanoparticles in biosensor development, they were used as separating tools in order to eliminate washing steps or attaching the sensing bodies on to the biosensor platform via magnetic field, there are increasing number of studies on the use of physicochemical properties of MNPs for sensing (Chemla, Grossman, Poon, McDermott, Stevens, Alper et al., 2000; Mody et al., 2010; Peng, Liang, Zhang & Qiu, 2013; Zhang, Carr et al., 2009).

Published study from Trahms's Lab in 1997 on a of superconducting quantum interference device (SQUID)-based detection system for the interaction of biomolecules is one of the first examples in which magnetic nanoparticles used as sensing body (Kötitz, Bunte, Weitschies & Trahms, 1997). This label free immunosensor bases mainly on the difference in magnetic relaxation time of magnetic nanoparticles before and after the reunion of target molecule (antigen) and the recognition element (antiobody) which is immobilized onto particle surface (Chemla et al., 2000; Titz, Matz, Drung, Hartwig, Groû & Ko, 1999).

In a more recent study, Baratella, Magro, Sinigaglia, Zboril, Salviulo and Vianello (2013) used maghemite nanoparticles attached on a carbon paste (CP) electrode as hydrogen peroxide electro-catalyst, and reported an oxidase based reagentless glucosensor device with a sensitivity of 45.85 nA $\mu M^{-1} cm^{-2}$, and a detection limit of 0.9 μM (Baratella et al., 2013). Same approach was conducted by incorporating magnetic nanoparticles to an exfoliated graphene oxide sheet with carboxyl-long-chains together with glucose oxide and poly[aniline-co-N-(1-one-butyric acid) aniline] (SPAnH) (Yang, Tjiu, Fan & Liu, 2013). In this study, glucose (they also detected hydrogen peroxide) was detected with sensitivity higher than that reported by Bratella et al. (1074.6 μA mM^{-1} cm^{-2}), however with a poorer linear detection range in mM level.

Esseghaier and Zourob were used magnetic particles with a handy procedure in order to detect HIV-1 protease which is essential for the HIV for the maintenance of its reproduciblity (Esseghaier & Zourob, 2012). They first built SEMs of peptide (HIV-1 protease substrate) conjugated magnetic particles constructed on gold electrode. Cleavage of the protease substrate by addition of HIV-1 protease under appropriate magnetic field, results in dissociation of the MNPs from the sensor surface (Figure 5). That kind of dissociation leads an electrochemical and optical signal change which is expected to be proportional to the target concentration. Authors claimed a detection limit as low as 100 pg HIV-1 protease /mL. This strategy could have potential with applicability to several other enzymes capable of cleaving, and can be extended to other types of biomolecules by designing engineered bridge molecules.

One other detection approach that we come across often is Giant-magnetoresistive (GMR) type sensors where the signal is measured due to the change in the resistivity of a material or a structure as a function of an external magnetic field. Due to claimed high sensitivity and quick response, GMR sensors have been utilized in many areas of science and technology including biosensor development with increased demand on materials showing high magnetoresistance. One of the examples of GMR sensors is the investigation of the uptake of macromolecules by cells. In such examples, magnetic beads left to reach with the cells, which previously grown on the biochip surface by sedimentation, and the progress of particle uptake, so that of phagocytosis, was monitored real-time by measuring the change in signal on a biochip system (Shoshi, Schotter, Schroeder, Milnera, Ertl, Charwat et al., 2012). In another example, Kim and Wang reported a magneto-nanosensor biochip for fungal detection by use of giant magnetoresistive (GMR) spin-valve sensor array in which biomarkers were successfully detected at pictogram levels (Dokyoon & Wang, 2012). These examples can be extended to genotyping of human hepatitis B virus (HBV), DNA and many other biomarkers (Li, Jing, Yao, Srinivasan, Xu, Xing et al., 2009; Xu, Yu, Han, Osterfeld, White, Pourmand & Wang, 2008). In addition to aforementioned analytical approaches, there are many other strategies aiming to make full potential of these unique materials (eg. fluxgate sensor (Baltag & Costandache, 1997; Ripka, 2003), Hall effect sensor (Volmer & Avram, 2013) and induction coil (Tumanski, 2007), exc.) in analytical chemistry. However, there are still more steps to be taken for understanding the magnetic behavior of them and finding better approaches for surface biofunctionalisation avoiding aggregations and loss of magnetic properties.

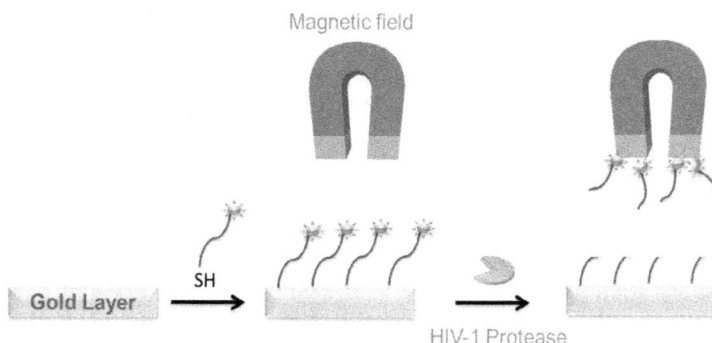

Figure 5. Mechanism of HIV-1 protease detection by magnetic nanoparticles.
Reprinted from Esseghaier and Zourob, 2012.

2.4. Carbon nanotubes

Carbon nanotubes (CNT) are carbon based tubular structures formed from two dimensional graphene sheets with nanometer-size diameters and micrometer lengths (Jorio, Dresselhaus, Dresselhaus, 2008). They hold excellent mechanical stiffness and chemical stability. Although, their history goes back to early 50s, CNTs became popular and noticed by scientific community after the paper published by Sumio Iijima in 1991 (Iijima, 1991). There are several techniques for the synthesis of carbon nanotubes with well-defined structure and dimensions (Dresselhaus, Dresselhaus & Avouris, 2001; Ren, 2007; Shanov, Yun & Schulz, 2006) as well as wall number (single wall, multi wall ext.) but demands on greener and facile synthesis methodologies are uprising. Since the first report focused on the catalytically properties of CNTs and their possible potential as biosensor element published in 1996 to present day, over 2000 articles and reviews were published according to Science Direct records in which over thousands of them were published in the last 5 years. This increased urge to the subject is due to their extraordinary mechanical features and conductivity in addition to their highly reactive surface with possibility to be functionalized with any molecule, making them ideal tools for the development of nanoelectronic devices, circuits and sensors (Balasubramanian & Burghard, 2006; Dresselhaus et al., 2001; Putzbach & Ronkainen, 2013). The surface functionalization of the CNTs with bacteria, aptamer, DNA, polymers, enzymes, metal nanoparticles and cells were demonstrated and used as multifunctional sensor platforms in Lab-on-Chip technologies (Bareket-Keren & Hanein, 2012; Chiariello, Miano & Maffucci, 2009; Choong, Milne & Teo, 2008; Yantzi & Yeow, 2005). Rius group published an aptosensor where they used high-affinity RNA aptamer that specifically binds to type IVB pili of Salmonella Typhi as recognition element (Zelada-Guillén, Riu, Düzgün & Rius, 2009). They immobilized aptamer onto the carbon nanotube, which horizontally attached to glassy carbon electrode through Π-Π stacking (Figure 6c). By detecting the charge changes due to a conformation change of aptamer as a result of its interaction with bacteria, they reported the possibility to detect extremely low concentrations of bacteria without cross reaction with other types of bacteria. A close strategy using glycosylated single walled CNTs (Figure 6c) was used to direct detection of secreted metabolites and biomolecules from cells (Sudibya, Ma, Dong, Ng, Li, Liu et al., 2009). These noninvasive methodologies for cell based detection can be extended to several other examples, such as genetic material, propteins, enzymes, aminoacids, cell proliferation, cancer biomarkers and viruses (Boero, Olivo,Carrara & De Micheli, 2012; Han, Doepke, Cho, Likodimos, de la Cruz, Back et al., 2013; Koehne et al., 2004; Lata, Batra, Kumar & Pundir, 2013; Merkoçi, 2009; Pandiaraj, Madasamy, Gollavilli, Balamurugan, Kotamraju, Rao et al., 2013; Yang et al., 2013; Zhang, Guo et al., 2009).

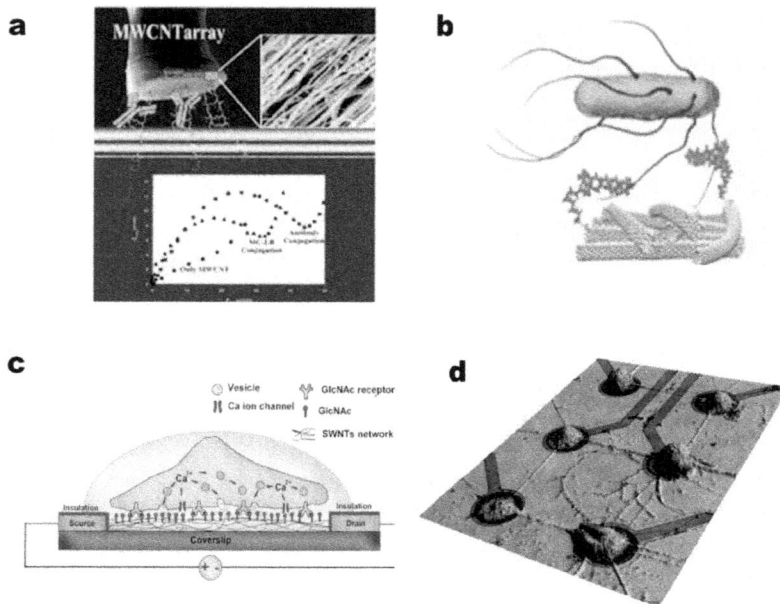

Figure 6. An assortment of different approaches for employing carbon nanotube as a part of biosensor for detection of various targets: a) Multiwalled-Carbon-Nanotube-Based biosensor for monitoring microcystin-LR in drinking water supplies (Han et al., 2013), b) aptamer based detection of bacteria on carbon nanotube embedded electrode (Zelada-Guillén et al., 2009), c) real time analysis of dynamic secretion from cells (Sudibya et al., 2009), and d) Neuro-gliacortical cell culture from embryonic rats grown on a carbon nanotube microelectrode array (Bareket-Keren & Hanein, 2012).

There are other approaches in which the shape of the CNTs comes forward. The 2D and 3D assembly of these high aspect ratio structures hollow inside, results in enormous high surface areas tasking account both the exterior and interior which allows not only immobilization of bioactive molecules on each CNTs but also forming highly catalytic surface allowing direct electron flow (Gomez-Gualdron, Burgos & Balbuena, 2011; Yang, Thordarson, Gooding, Ringer & Braet, 2010). AMES Research Center, NASA, researchers reported a label free DNA microchip built up of vertically aligned multiwalled carbon nanotubes (MWCNTs) by wafer-scale plasma enhanced chemical vapor deposition (PECVD) method (Koehne et al., 2004). The probe molecules (antibody, nucleic acids vs.) were attached to the tips of MWCNTs and analyte concentration was directly measured from oxidized guanine signal with several fold higher sensitivity, low detection limit and with multiplex property.

High aspect ratio CNTs also found a lot of use in the formation of composites where they have been employed for the growth of various kinds of metal and metal oxide coated or decorated nanotube structures and composites (Ajayan, Stephan, Redlich & Colliex, 1995; Chastel, Flahaut, Peigney & Rousset, 2000; Chen, Hu, Shao, Li & Wang, 2009; Harris, 2004). These catalytically or optically activated sensor surfaces demonstrated high sensitivity and selectivity for the label free detection of biomolecules. (Baro, Nayak, Baby & Ramaprabhu, 2013; Jiang et al., 2011; Kumar, Mehdipour & Ostrikov, 2013)

Despite their huge potentials in sensor technology, the non-ideal (or even cytotoxic) interface between CNTs and the living cells, asbestos like pathogenicity, limits their application to biological systems and wider use (Ali-Boucetta, Nunes, Sainz, Herrero, Tian, Prato et al., 2013; Osmond-McLeod, Poland, Murphy, Waddington, Morris, Hawkins et al., 2011; Poland, Duffin, Kinloch, Maynard, Wallace, Seaton et al., 2008), thus, new coating materials and strategies in order to overcome biocompatibility issues as well as new synthesis techniques to increase deposition homogeneity and to make the *in situ* nanotube growth easier are strongly required.

3. Conclusions and Future Aspects

Today scientists are capable of developing sensing systems to detect biomarkers, essential proteins, genetic material, and so on, with the help of nanotechnology. Nanotechnology holds a great potential for imaging, diagnosis and therapy, where different nanostructures from different material could be used as labels, signal enhancers, catalyst and as tools to build sensor platforms. The final aim is to obtain efficient techniques for improved real-time molecular imaging, monitoring drug trafficking, as well as development of Lab-on-a-Chip devices capable of detecting diseases and report real-time. Although, still the use of them in clinical requires extensive effort, there are several ongoing projects worldwide aiming to understand the basis of nanoworld. The forthcoming years would see their potential applications in building Lab-on-Chip system for home-based point-of-care diagnosis. However, how much harmful are those nanoparticles and nanostructures to the living organisms and environment is still a major question to be concerned and should be resolved before they engaged to in bio-related areas.

References

Ajayan, P.M., Stephan, O., Redlich, P., & Colliex, C. (1995). Carbon Nanotubes as Removable Templates for Metal-Oxide Nanocomposites and Nanostructures. *Nature, 375(6532),* 564-567. http://dx.doi.org/10.1038/375564a0

Algar, W.R., Malanoski, A.P., Susumu, K., Stewart, M.H., Hildebrandt, N., & Medintz, I.L. (2012). Multiplexed Tracking of Protease Activity Using a Single Color of Quantum Dot Vector and a Time-Gated Förster Resonance Energy Transfer Relay. *Analytical chemistry*, 84(22), 10136-10146. http://www.ncbi.nlm.nih.gov/pubmed/23128345 (Last access date: January 2014). http://dx.doi.org/10.1021/ac3028068

Ali-Boucetta, H., Nunes, A., Sainz, R., Herrero, M.A., Tian, B., Prato, M., et al. (2013). Asbestos-like pathogenicity of long carbon nanotubes alleviated by chemical functionalization. *Angewandte Chemie, 52(8),* 2274-2278. http://www.ncbi.nlm.nih.gov/pubmed/23319294 (Last access date: June 2013).

Alvarez-Puebla, R.A., Zubarev, E.R., Kotov, N.A., & Liz-Marzán, L.M. (2012). Self-assembled nanorod supercrystals for ultrasensitive SERS diagnostics. *Nano Today, 7(1),* 6-9. http://linkinghub.elsevier.com/retrieve/pii/S1748013211001332 (Last access date: June 2013). http://dx.doi.org/10.1016/j.nantod.2011.11.001

Alwi, R., Telenkov, S., Mandelis, A., Leshuk, T., Gu, F., Oladepo, S., et al. (2012). Silica-coated super paramagnetic iron oxide nanoparticles (SPION) as biocompatible contrast agent in biomedical photoacoustics. *Biomedical Optics Express, 3(10),* 2500-2509. http://www.pubmedcentral.nih.gov/articlerender.fcgi?artid=3470002&tool=pmcentrez&rendertype=abstract
http://dx.doi.org/10.1364/BOE.3.002500

Amna, T., Van Ba, H., Vaseem, M., Hassan, M.S., Khil, M.S., Hahn, Y.B., et al. (2013). Apoptosis induced by copper oxide quantum dots in cultured C2C12 cells via caspase 3 and caspase 7: a study on cytotoxicity assessment. *Applied Microbiology and Biotechnology, 97(12),* 5545-5553. http://www.ncbi.nlm.nih.gov/pubmed/23467821 (Last access date: June 2013).
http://dx.doi.org/10.1007/s00253-013-4724-1

Ankri, R., Peretz, V., Motiei, M., Popovtzer, R., & Fixler, D. (2012). A new method for cancer detection based on diffusion reflection measurements of targeted gold nanorods. *International Journal of Nanomedicine, 7,* 449-455. http://www.pubmedcentral.nih.gov/articlerender.fcgi?artid=3273979&tool=pmcentrez&rendertype=abstract

Asokan, S., Krueger, K.M., Alkhawaldeh, A., Carreon, A.R., Mu, Z., Colvin V.L., et al. (2005). The use of heat transfer fluids in the synthesis of high-quality CdSe quantum dots, core/shell quantum dots, and quantum rods. *Nanotechnology, 16(10),* 2000-2011. http://stacks.iop.org/0957-4484/16/i=10/a=004?key=crossref.df474413a7fa832dc1e25fd8996d4182 http://dx.doi.org/10.1088/0957-4484/16/10/004

Balasubramanian, K., & Burghard, M. (2006). Biosensors based on carbon nanotubes. *Analytical and Bioanalytical Chemistry, 385(3),* 452-468. http://www.ncbi.nlm.nih.gov/pubmed/16568294 (Last access date: May 2013). http://dx.doi.org/10.1007/s00216-006-0314-8

Baltag, O., & Costandache, D. (1997). Sensor with ferrofluid for magnetic measurements. *Proceedings of 20th Biennial Conference on Precision Electromagnetic Measurements, 46(2),* 629-631. http://ieeexplore.ieee.org/lpdocs/epic03/wrapper.htm?arnumber=571941

Bang, J.H., & Kamat, P.V. (2009). Quantum dot sensitized solar cells. A tale of two semiconductor nanocrystals: CdSe and CdTe. *ACS Nano, 3(6),* 1467-1476. http://www.ncbi.nlm.nih.gov/pubmed/19435373 http://dx.doi.org/10.1021/nn900324q

Baratella, D., Magro, M., Sinigaglia, G., Zboril, R., Salviulo, G., & Vianello, F. (2013). A glucose biosensor based on surface active maghemite nanoparticles. *Biosensors Bioelectronics, 45,* 13-18. http://www.ncbi.nlm.nih.gov/pubmed/23454337 http://dx.doi.org/10.1016/j.bios.2013.01.043

Bareket-Keren, L., & Hanein, Y. (2012). Carbon nanotube-based multi electrode arrays for neuronal interfacing: progress and prospects. *Frontiers in Neural Circuits, 6(January),* 122. http://www.pubmedcentral.nih.gov/articlerender.fcgi?artid=3540767&tool=pmcentrez&rendertype=abstract (Last access date: May 2013).

Baro, M., Nayak, P., Baby, T.T., & Ramaprabhu, S. (2013). Green approach for the large-scale synthesis of metal/metal oxide nanoparticle decorated multiwalled carbon nanotubes. *Journal of Materials Chemistry A, 1(3),* 482. http://xlink.rsc.org/?DOI=c2ta00483f (Last access date: June 2013). http://dx.doi.org/10.1039/c2ta00483f

Beqa, L., Singh, A.K., Fan, Z., Senapati, D., & Ray, P.C. (2011). Chemically attached gold nanoparticle-carbon nanotube hybrids for highly sensitive SERS substrate. *Chemical Physics Letters, 512(4-6),* 237-242. http://linkinghub.elsevier.com/retrieve/pii/S0009261411008530 (Last access date: June 2013). http://dx.doi.org/10.1016/j.cplett.2011.07.037

Boero, C., Olivo, J., Carrara, S., & De Micheli, G. (2012). A Self-Contained System With CNTs-Based Biosensors for Cell Culture Monitoring. *IEEE Journal on Emerging and Selected Topics in Circuits and Systems, 2(4),* 658-671. http://ieeexplore.ieee.org/lpdocs/epic03/wrapper.htm?arnumber=6365265 http://dx.doi.org/10.1109/JETCAS.2012.2223592

Burda, C., Chen, X., Narayanan, R., & El-Sayed, M.A. (2005). Chemistry and properties of nanocrystals of different shapes. *Chemical Reviews, 105,* 1025-1102. http://www.ncbi.nlm.nih.gov/pubmed/15826010 http://dx.doi.org/10.1021/cr030063a

Byun, K.M. (2010). Development of Nanostructured Plasmonic Substrates for Enhanced Optical Biosensing. *Journal of the Optical Society of Korea, 14(2),* 65-76. http://koreascience.or.kr/journal/view.jsp?kj=E1OSAB&py=2010&vnc=v14n2&sp=65 (Last access date: June 2013). http://dx.doi.org/10.3807/JOSK.2010.14.2.065

Chastel, F., Flahaut, E., Peigney, A., & Rousset, A. (2000). Carbon nanotube-metal-oxide nanocomposites: microstructure, electrical conductivity and mechanical properties. *Acta Materialia, 48(14),* 3803-3812. http://linkinghub.elsevier.com/retrieve/pii/S1359645400001476 http://dx.doi.org/10.1016/S1359-6454(00)00147-6

Chemla, Y.R., Grossman, H.L., Poon, Y., McDermott, R., Stevens, R., Alper, M.D., et al. (2000). Ultrasensitive magnetic biosensor for homogeneous immunoassay. *Proceedings of the National Academy of Sciences of the United States of America, 97(26),* 14268-14272. http://www.pubmedcentral.nih.gov/articlerender.fcgi?artid=18907&tool=pmcentrez&rendertype=abstract http://dx.doi.org/10.1073/pnas.97.26.14268

Chen, C., Hu, J., Shao, D., Li, J., & Wang, X. (2009). Adsorption behavior of multiwall carbon nanotube/iron oxide magnetic composites for Ni(II) and Sr(II). *Journal of Hazardous Materials, 164(2-3),* 923-928. http://www.ncbi.nlm.nih.gov/pubmed/18842337 http://dx.doi.org/10.1016/j.jhazmat.2008.08.089

Cheng, H.W., Huan, S.Y., & Yu, R.Q. (2012). Nanoparticle-based substrates for surface-enhanced Raman scattering detection of bacterial spores. *The Analyst, 137(16),* 3601-3608. http://www.ncbi.nlm.nih.gov/pubmed/22745931 http://dx.doi.org/10.1039/c2an35448a

Chiariello, A.G., Miano, G., & Maffucci, A. (2009). Carbon nanotube bundles as nanoscale chip to package interconnects. *9th IEEE Conference on Nanotechnology. IEEE-NANO 2009, 8,* 58-61. http://www.ncbi.nlm.nih.gov/entrez/query.fcgi? db=pubmed&cmd=Retrieve&dopt=AbstractPlus&list_uids=5394569

Choi, Y., Ho, N.H., & Tung, C.H. (2007). Sensing phosphatase activity by using gold nanoparticles. *Angewandte Chemie (International ed. in English), 46(5),* 707-709. http://www.ncbi.nlm.nih.gov/pubmed/17143915 (Last access date: May 2013). http://dx.doi.org/10.1002/anie.200603735

Choong, C.L., Milne, W.I., & Teo, K.B. (2008). Review: carbon nanotube for microfluidic lab-on-a-chip application. *International Journal of Material Forming, 1(2),* 117-125. http://www.springerlink.com/index/10.1007/s12289-008-0379-3 http://dx.doi.org/10.1007/s12289-008-0379-3

Costa, M.M.D., De la Escosura-Muñiz, A., & Merkoçi, A. (2010). Electrochemical quantification of gold nanoparticles based on their catalytic properties toward hydrogen formation: Application in magnetoimmunoassays. *Electrochemistry Communications, 12(11),* 1501-1504. http://linkinghub.elsevier.com/retrieve/pii/S1388248110003644 (Last access date: June 2013). http://dx.doi.org/10.1016/j.elecom.2010.08.018

D'Agata, R., & Spoto, G. (2013). Surface plasmon resonance imaging for nucleic acid detection. *Analytical and Bioanalytical Chemistry, 405(2-3),* 573-584. http://www.ncbi.nlm.nih.gov/pubmed/23187826 (Last access date: June 2013). http://dx.doi.org/10.1007/s00216-012-6563-9

Dabbousi, B.O., Rodriguez-Viejo, J., Mikulec, F.V., Heine, J.R., Mattoussi, H., Ober, R., et al. (1997). (CdSe)ZnS Core–Shell Quantum Dots: Synthesis and Characterization of a Size Series of Highly Luminescent Nanocrystallites. *The Journal of Physical Chemistry B, 101(46),* 9463-9475. http://pubs.acs.org/doi/abs/10.1021/jp971091y http://dx.doi.org/10.1021/jp971091y

Dahnovsky, Y.I., Krevchik, V.D., Semenov, M.B., Yamamoto, K., Zhukovsky, V.C., Aringazin, A.K., et al. (2005). Dissipative tunneling in structures with quantum dots and quantum molecules. *Physics,* 18. http://arxiv.org/abs/cond-mat/0509119

De La Escosura-Muñiz, A., & Merkoçi, A. (2011). A nanochannel/nanoparticle-based filtering and sensing platform for direct detection of a cancer biomarker in blood. *Small Weinheim an der Bergstrasse Germany, 7(5),* 675-682. http://www.ncbi.nlm.nih.gov/pubmed/21294272 http://dx.doi.org/10.1002/smll.201002349

Dokyoon, K., & Wang, S.X. (2012). A Magneto-Nanosensor Immunoassay for Sensitive Detection of Aspergillus Fumigatus Allergen. *IEEE Transactions on Magnetics, 48(11),* 3266-3268. http://dx.doi.org/10.1109/TMAG.2012.2195163

Dresselhaus, M.S., Dresselhaus, G., & Avouris, P. (2001). Carbon nanotubes: synthesis, structure, properties, and applications. 1 *Journal of Physical Chemistry*, 6-53. Dresselhaus, M.S. Dresselhaus, G., & Avouris, P (Eds.). Springer. http://books.google.com/books?hl=en&lr=&id=dkvDhZJnafgC&oi=fnd&pg=PA1&dq=Carbon+Nanotubes+Synthesis,+Structure,+Properties,+and+Applications&ots=G4yzOoZmDu&sig=mBJEM2H2cSiG9GkUX_apROjiHNs

Esseghaier, C., Ng, A., & Zourob, M. (2012). A novel assay for rapid HIV-1 protease detection using optical sensors and magnetic carriers. *Photonics North 2012*, International Society for Optics and Photonics, 841209-841209-6. http://proceedings.spiedigitallibrary.org/proceeding.aspx?articleid=1387269 (Last access date: June 2013).

Fang, J., Liu. S., & Li, Z. (2011). Polyhedral silver mesocages for single particle surface-enhanced Raman scattering-based biosensor. *Biomaterials, 32(21),* 4877-4884. http://www.ncbi.nlm.nih.gov/pubmed/21492933 (Last access date: June 2013). http://dx.doi.org/10.1016/j.biomaterials.2011.03.029

Floris, A., Ardu, A., Musinu, A., Piccaluga, G., Fadda, A.M., Sinico, C. et al. (2011). SPION@liposomes hybrid nanoarchitectures with high density SPION association. *Soft Matter, 7(13),* 6239-6247. http://dx.doi.org/10.1039/c1sm05059a

Frantsuzov, P.A., & Marcus, R.A. (2005). Explanation of quantum dot blinking without long-lived trap hypothesis. *Physical Review B. 72(15),* 1-10. http://arxiv.org/abs/cond-mat/0505604 http://dx.doi.org/10.1103/PhysRevB.72.155321

Gao, Y., Lam, A.W., & Chan, W.C. (2013). Automating Quantum Dot Barcode Assays Using Microfluidics and Magnetism for the Development of a Point-of-Care Device. *ACS Applied Materials & Interfaces, 5(8),* 2853-2860. http://www.ncbi.nlm.nih.gov/pubmed/23438061 http://dx.doi.org/10.1021/am302633h

Genç, R., Ortiz, M., & O'Sullivan, C.K. (2009). Curvature-Tuned Preparation of Nanoliposomes. *Langmuir, 25(21),* 12604-12613. http://www.ncbi.nlm.nih.gov/pubmed/19856992 http://dx.doi.org/10.1021/la901789h

Goldhawk, D.E., Rohani, R., Sengupta, A., Gelman, N., & Prato, F.S. (2012). Using the magnetosome to model effective gene-based contrast for magnetic resonance imaging. *Wiley Interdisciplinary Reviews Nanomedicine and nanobiotechnology, 4(4),* 378-388. http://www.ncbi.nlm.nih.gov/pubmed/22407727 http://dx.doi.org/10.1002/wnan.1165

Gomez-Gualdron, D.A., Burgos, J.C, Yu, J., & Balbuena, P.B. (2011). Carbon Nanotubes: Engineering Biomedical Applications. *Progress in Molecular Biology and Translational Science, 104,* 175-245. http://dx.doi.org/10.1016/B978-0-12-416020-0.00005-X

Guerrini, L., & Graham, D. (2012). Molecularly-mediated assemblies of plasmonic nanoparticles for Surface-Enhanced Raman Spectroscopy applications. *Chemical Society Reviews, 41(21),* 7085-7107. http://www.ncbi.nlm.nih.gov/pubmed/22833008 (Last access date: June 2013).

Gunawidjaja, R., Kharlampieva, E., Choi, I., & Tsukruk, V.V. (2009). Bimetallic Nanostructures as Active Raman Markers : Gold-Nanoparticle Assembly on 1D and 2D Silver Nanostructure Surfaces. *Materials Science, (21),* 2460-2466.

Guo, S., & Wang, E. (2007). Synthesis and electrochemical applications of gold nanoparticles. *Analytica Chimica Acta, 598(2),* 181-192. http://www.ncbi.nlm.nih.gov/pubmed/17719891 (Last access date: May 2013). http://dx.doi.org/10.1016/j.aca.2007.07.054

Gupta, A.K., & Gupta, M. (2005). Synthesis and surface engineering of iron oxide nanoparticles for biomedical applications. *Biomaterials, 26(18),* 3995-4021. http://www.ncbi.nlm.nih.gov/pubmed/15626447 (Last access date: March 2012). http://dx.doi.org/10.1016/j.biomaterials.2004.10.012

Han, C., Doepke, A., Cho, W., Likodimos, V., de la Cruz, A.A., Back, T., et al. (2013). A Multiwalled-Carbon-Nanotube-Based Biosensor for Monitoring Microcystin-LR in Sources of Drinking Water Supplies. *Advanced Functional Materials, 23(14),* 1807-1816. http://doi.wiley.com/10.1002/adfm.201201920 (Last access date: June 2013). http://dx.doi.org/10.1002/adfm.201201920

Harris, P.J.F. (2004). Carbon nanotube composites. *International Materials Reviews, 49(1),* 31-43. http://www.ingentaselect.com/rpsv/cgi-bin/cgi?ini=xref&body=linker&reqdoi=10.1179/095066004225010505 http://dx.doi.org/10.1179/095066004225010505

Hartgerink, J.D., Beniash, E., & Stupp, S.I. (2001). Self-Assembly and Mineralization of Peptide-Amphiphile Nanofibers. *Science,* 1684. http://dx.doi.org/10.1126/science.1063187

Hill, H.D., & Mirkin, C.A. (2006). The bio-barcode assay for the detection of protein and nucleic acid targets using DTT-induced ligand exchange. *Nature Protocols, 1(1),* 324-336. http://www.ncbi.nlm.nih.gov/pubmed/17406253 http://dx.doi.org/10.1038/nprot.2006.51

Himmelhaus, M., & Takei, H. (2000). Cap-shaped gold nanoparticles for an optical biosensor. *Sensors and Actuators B: Chemical, 63(1-2),* 24-30. http://linkinghub.elsevier.com/retrieve/pii/S0925400599003937 http://dx.doi.org/10.1016/S0925-4005(99)00393-7

Iijima, S. (1991). Helical microtubules of graphitic carbon. *Nature,* 354(6348), 56-58. http://www.nature.com/doifinder/10.1038/354056a0 http://dx.doi.org/10.1038/354056a0

Jiang, W.F., Zhang, Y.F., Wang, Y.S., Xu, L., & Li, S.J. (2011). SERS activity of Au nanoparticles coated on an array of carbon nanotube nested into silicon nanoporous pillar. *Applied Surface Science, 258(5),* 1662-1665. http://linkinghub.elsevier.com/retrieve/pii/S0169433211015170 (Last access date: June 2013). http://dx.doi.org/10.1016/j.apsusc.2011.09.114

Jianrong, C., Yuqing, M., Nongyue, H., Xiaohua, W., & Sijiao, L. (2004). Nanotechnology and biosensors. *Biotechnology Advances, 22(7),* 505-518. http://www.ncbi.nlm.nih.gov/pubmed/15262314 (Last access date: May 2013). http://dx.doi.org/10.1016/j.biotechadv.2004.03.004

Jorio, A.; Dresselhaus, G., & Dresselhaus, M.S. (Eds.). (2008). *Carbon Nanotubes: Advanced Topics in the Synthesis, Structure, Properties and Applications, 111*. Springer Netherlands.

Kattumenu, R., Lee, C.H., Tian, L., McConney, M.E., & Singamaneni, S. (2011). Nanorod decorated nanowires as highly efficient SERS-active hybrids. *Journal of Materials Chemistry, 21(39),* 15218. http://xlink.rsc.org/?DOI=c1jm12426a (Last access date: June 2013).
http://dx.doi.org/10.1039/c1jm12426a

Khoury, C.G., & Vo-Dinh, T. (2008). Gold Nanostars For Surface-Enhanced Raman Scattering : Synthesis, Characterization and Optimization. *Growth (Lakeland),* 18849-18859.

Kim, D.K., Mikhaylova, M., Wang, F.H., Kehr, J, Bjelke, B., Zhang, Y., et al. (2003). Starch-coated superparamagnetic nanoparticles as MR contrast agents. *Chemistry of Materials, 15(23),* 4343-4351. http://pubs.acs.org/doi/abs/10.1021/cm031104m http://dx.doi.org/10.1021/cm031104m

Kim, D.K., Yoo, S.M., Park, T.J, Yoshikawa, H., Tamiya, E., Park, J.Y., et al. (2011). Plasmonic Properties of the Multispot Copper-Capped Nanoparticle Array Chip and Its Application to Optical Biosensors for Pathogen Detection of Multiplex DNAs. *Anal. Chem., 83(16),* 6215-6222. http://www.ncbi.nlm.nih.gov/pubmed/21714496. http://dx.doi.org/10.1021/ac2007762

Kim, J.H., Hiraiwa, M., Lee, H.B., & Lee, K.H. (2013). Electrolyte-free amperometric immunosensor using a dendritic nanotip. *RSC Advances, 3,* 4281-4287. http://dx.doi.org/10.1039/c3ra40252b

Kneipp, J., Kneipp, H., McLaughlin, M., Brown, D., & Kneipp, K. (2006). In vivo molecular probing of cellular compartments with gold nanoparticles and nanoaggregates. *Nano Letters, 6(10),* 2225-2231. http://www.ncbi.rlm.nih.gov/pubmed/17034088 http://dx.doi.org/10.1021/nl061517x

Knight, V.B., & Serrano, E.E. (2006). Tissue and Species Differences in the Application of Quantum Dots as Probes for Biomolecular Targets in the Inner Ear and Kidney. *IEEE Transactions on NanoBioscience, 5(4),* 251-262. http://www.ncbi.nlm.nih.gov/pubmed/17181024 http://dx.doi.org/10.1109/TNB.2006.886551

Koehne, J.E., Chen, H., Cassell, A.M., Ye, Q., Han, J., Meyyappan, M., et al. (2004). Miniaturized multiplex label-free electronic chip for rapid nucleic acid analysis based on carbon nanotube nanoelectrode arrays. *Clinical Chemistry, 50(10),* 1886-1893. http://www.ncbi.nlm.nih.gov/pubmed/15319319 (Last access date: June 2013). http://dx.doi.org/10.1373/clinchem.2004.036285

Kötitz, R., Bunte, T., Weitschies, W., & Trahms, L. (1997). Superconducting quantum interference device-based magnetic nanoparticle relaxation measurement as a novel tool for the binding specific detection of biological binding reactions. *Journal of Applied Physics, 81(8),* 4317. http://dx.doi.org/10.1063/1.364754

Kumar, S., Mehdipour, H., & Ostrikov, K. K. (2013). Plasma-enabled graded nanotube biosensing arrays on a Si nanodevice platform: catalyst-free integration and in situ detection of nucleation events. *Advanced Materials (Deerfield Beach, Fla.), 25(1),* 69-74. http://www.ncbi.nlm.nih.gov/pubmed/23108975 (Last access date: June 2013). http://dx.doi.org/10.1002/adma.201203163

Lata, S., Batra, B., Kumar, P., & Pundir, C.S. (2013). Construction of an amperometric d-amino acid biosensor based on d-amino acid oxidase/carboxylated mutliwalled carbon nanotube/copper nanoparticles/polyalinine modified gold electrode. *Analytical Biochemistry, 437(1),* 1-9. http://www.ncbi.nlm.nih.gov/pubmed/23399389 (Last access date: May 2013). http://dx.doi.org/10.1016/j.ab.2013.01.030

Lee, S.H., Sung, J.H., & Park, T.H. (2011). Nanomaterial-Based Biosensor as an Emerging Tool for Biomedical Applications. *Annals of Biomedical Engineering, 40(6),* 1384-1397. http://www.ncbi.nlm.nih.gov/pubmed/22065202 http://dx.doi.org/10.1007/s10439-011-0457-4

Lee, S.F., & Osborne, M.A. (2009). Brightening, Blinking, Bluing and Bleaching in the Life of a Quantum Dot: Friend or Foe?. *Chemphyschem A European Journal Of Chemical Physics And Physical Chemistry, 10(13),* 2174-2191. http://dx.doi.org/10.1002/cphc.200900200

Li, T., Hu, W., & Zhu, D. (2010). Nanogap electrodes. *Advanced materials Deerfield Beach Fla, 22(2),* 286-300. http://www.ncbi.nlm.nih.gov/pubmed/20217688 http://dx.doi.org/10.1002/adma.200900864

Li, Y., Jing, Y., Yao, X., Srinivasan, B., Xu, Y., Xing C., et al. (2009). Biomarkers identification and detection based on GMR sensor and sub 13 nm magnetic nanoparticles. *Conference Proceedings of the International Conference of IEEE Engineering in Medicine and Biology Society* 2009, 5432-5435.

Lin, Y.C., Bai, J., & Huang, Y. (2009). Self-aligned nanolithography in a nanogap. *Nano Letters, 9(6),* 2234-2238. http://www.ncbi.nlm.nih.gov/pubmed/19413343 http://dx.doi.org/10.1021/nl9000597

Liu, G., Wang, J., & Lin, Y. (2006). Nanoparticle Labels/Electrochemical Immunosensor for Detection of Biomarkers. *NTSINanotech, 2,* 192-195.

Lowe, S.B., Dick, J.A., Cohen, B.E., & Stevens, M.M. (2011). Multiplex Sensing of Protease and Kinase Enzyme Activity via Orthogonal Coupling of Quantum Dot-Peptide Conjugates. *ACS Nano, 6(1),* 851-857. http://www.ncbi.nlm.nih.gov/pubmed/22148227 (Las access date: January 2014). http://dx.doi.org/10.1021/nn204361s

Mahdihassan, S. (1984). Outline of the beginnings of alchemy and its antecedents. *The American Journal Of Chinese Medicine, 12(1-4),* 32-42. http://www.ncbi.nlm.nih.gov/entrez/query.fcgi?cmd=Retrieve&db=PubMed&dopt=Citation&list_uids=6388307 http://dx.doi.org/10.1142/S0192415X84000039

Matsumura, Y., Yoshikata, K., Kunisaki, S.I., & Tsuchido, T. (2003). Mode of Bactericidal Action of Silver Zeolite and Its Comparison with That of Silver Nitrate. *Society, 69(7),* 4278-4281. http://aem.asm.org/content/69/7/4278.short

Merkoçi, A. (Ed.) (2009). *Structure Biosensing Using Nanomaterials*. John Wiley & Sons, Inc. http://doi.wiley.com/10.1002/9780470447734 http://dx.doi.org/10.1002/9780470447734

Merkoçi, A. (2013). Nanoparticles Based Electroanalysis in Diagnostics Applications. *Electroanalysis, 25(1),* 15-27. http://doi.wiley.com/10.1002/elan.201200476 (Last access date: June 2013). http://dx.doi.org/10.1002/elan.201200476

Michalet, X., Pinaud, F.F., Bentolila, L.A., Tsay, J.M., Doose, S., Li, J.J., et al. (2005). Quantum dots for live cells, in vivo imaging, and diagnostics. *Science, 307(5709),* 538-544. http://www.pubmedcentral.nih.gov/articlerender.fcgi?artid=1201471&tool=pmcentrez&rendertype=abstract (Last access date: May 2013).

Mikhaylova, M., Kim, D. K., Bobrysheva, N., Osmolowsky, M., Semenov, V., Tsakalakos, T., et al. (2004). Superparamagnetism of magnetite nanoparticles: dependence on surface modification. *Langmuir The Acs Journal Of Surfaces And Colloids, 20(6),* 2472-2477. http://www.ncbi.nlm.nih.gov/pubmed/15835712 http://dx.doi.org/10.1021/la035648e

Mody, V.V., Siwale, R., Singh, A., & Mody, H.R. (2010). Introduction to metallic nanoparticles. *Journal of Pharmacy and Bioallied Sciences, 2(4),* 282-289. http://www.pubmedcentral.nih.gov/articlerender.fcgi?artid=2996072&tool=pmcentrez&rendertype=abstract http://dx.doi.org/10.4103/0975-7406.72127

Nam, J.M., Wise, A.R., & Groves, J.T. (2005). Colorimetric bio-barcode amplification assay for cytokines. *Analytical Chemistry, 77(21),* 6985-6988. http://www.ncbi.nlm.nih.gov/pubmed/16255599 http://dx.doi.org/10.1021/ac0513764

Omidfar, K., Khorsand, F., & Azizi, M.D. (2013). New analytical applications of gold nanoparticles as label in antibody based sensors. *Biosensors & Bioelectronics, 43,* 336-347. http://www.ncbi.nlm.nih.gov/pubmed/23356999 (Last access date: June 2013). http://dx.doi.org/10.1016/j.bios.2012.12.045

Osmond-McLeod, M.J., Poland, C.A., Murphy, F., Waddington, L., Morris, H., Hawkins, S.C., et al. (2011). Durability and inflammogenic impact of carbon nanotubes compared with asbestos fibres. *Particle and Fibre Toxicology, 8(1),* 15. http://www.pubmedcentral.nih.gov/articlerender.fcgi?artid=3126712&tool=pmcentrez&rendertype=abstract http://dx.doi.org/10.1186/1743-8977-8-15

Pardiaraj, M., Madasamy, T., Gollavilli, P.N., Balamurugan, M., Kotamraju, S., Rao, V.K., et al. (2013). Nanomaterial-based electrochemical biosensors for cytochrome c using cytochrome c reductase. *Bioelectrochemistry, 91,* 1-7. http://www.ncbi.nlm.nih.gov/pubmed/23220491 (Last access date: May 2013). http://dx.doi.org/10.1016/j.bioelechem.2012.09.004

Patolsky, F., Weizmann, Y., Lioubashevski, O., & Willner, I. (2002). Au-Nanoparticle Nanowires Based on DNA and polylysine templates. Angerwandte Chemie (International ed. In english) 41(13), 2323-2327. http://www.ncbi.nlm.nih.gov/pubmed/12203580 (Last access date: January 2014). http://dx.doi.org/10.1002/1521-3773(20020703)41:13<2323::AID-ANIE2323>3.0.CO;2-H

Peng, H.P., Liang, R.P., Zhang, L., & Qiu, J.D. (2013). Facile preparation of novel core-shell enzyme-Au-polydopamine-Fe_3O_4 magnetic bionanoparticles for glucosesensor. *Biosensors & Bioelectronics, 42,* 293-299. http://www.ncbi.nlm.nih.gov/pubmed/23208101 (Last access date: June 2013). http://dx.doi.org/10.1016/j.bios.2012.10.074

Poland, C.A., Duffin, R., Kinloch, I., Maynard, A., Wallace, W.A., Seaton, A., et al. (2008). Carbon nanotubes introduced into the abdominal cavity of mice show asbestos-like pathogenicity in a pilot study. *Nature Nanotechnology, 3(7),* 423-428. http://www.ncbi.nlm.nih.gov/pubmed/18654567 http://dx.doi.org/10.1038/nnano.2008.111

Pustovit, V.N., & Shahbazyan, T.V. (2006). Finite-size effects in surface-enhanced Raman scattering in noble-metal nanoparticles: a semiclassical approach. *Journal of the Optical Society of America A, 23(6),* 1369-1374. http://www.ncbi.nlm.nih.gov/pubmed/16715155 http://dx.doi.org/10.1364/JOSAA.23.001369

Putzbach, W., & Ronkainen, N.J. (2013). Immobilization techniques in the fabrication of nanomaterial-based electrochemical biosensors: a review. *Sensors, 13(4),* 4811-4840. http://www.ncbi.nlm.nih.gov/pubmed/23580051 (Last access date: May 2013). http://dx.doi.org/10.3390/s130404811

Ranjan, M., & Facsko, S. (2012). Anisotropic surface enhanced Raman scattering in nanoparticle and nanowire arrays. *Nanotechnology, 23(48),* 485307. http://www.ncbi.nlm.nih.gov/pubmed/23128982 (Last access date: June 2013). http://dx.doi.org/10.1088/0957-4484/23/48/485307

Ren, Z.F. (2007). Nanotube synthesis-Cloning carbon. *Nature Nanotechnology, 2(1),* 17-18. http://dx.doi.org/10.1038/nnano.2006.192

Resch-Genger, U., Grabolle, M., Cavaliere-Jaricot, S., Nitschke, R., & Nann, T. (2008). Quantum dots versus organic dyes as fluorescent labels. *Nature Methods, 5(9).* http://dx.doi.org/10.1038/nmeth.1248

Ripka, P. (2003). Advances in fluxgate sensors. *Sensors and Actuators A: Physical, 106(1-3),* 8-14. http://linkinghub.elsevier.com/retrieve/pii/S0924424703000943 http://dx.doi.org/10.1016/S0924-4247(03)00094-3

Rissin, D.M., Kan C.W., Song, L., Rivnak, A.J., Fishburn, M.W., Shao, Q., et al. (2013). Multiplexed single molecule immunoassays. *Lab on a Chip.* http://xlink.rsc.org/?DOI=c3lc50416f (Last access date: May 2013).

Rumyantseva, A., Kostcheev, S., Adam, P.M., Gaponenko, S.V., Vaschenko, S.V., Kulakovich, O.S., et al. (2013). Nonresonant Surface-Enhanced Raman Scattering of ZnO Quantum Dots with Au and Ag Nanoparticles. *ACS Nano, 7(4),* 3420-3426.
http://www.ncbi.nlm.nih.gov/pubmed/23464800 http://dx.doi.org/10.1021/nn400307a

Saha, K., Agasti, S.S., Kim, C., Li, X., & Rotello, V.M. (2012). Gold nanoparticles in chemical and biological sensing. *Chemical Reviews, 112(5),* 2739-2379.
http://www.ncbi.nlm.nih.gov/pubmed/22295941 http://dx.doi.org/10.1021/cr2001178

Sato, K., Hosokawa, K. & Maeda, M. (2007). Colorimetric Biosensors Based on DNA-nanoparticle Conjugates. *Analytical Chemistry, 23(1),* 17-20.

Sawant, R.M., Sawant, R.R., Gultepe, E., Nagesha, D., Papahadjopoulos-Sternberg, B., Sridhar, S., et al. (2009). Nanosized cancer cell-targeted polymeric immunomicelles loaded with superparamagnetic iron oxide nanoparticles. *Journal of Nanoparticle Research, 11(7),* 1777-1785. http://www.springerlink.com/content/p2m33612178q02n0/ http://dx.doi.org/10.1007/s11051-009-9611-4

Schütz, M., Steinigeweg, D., Salehi, M., Kömpe, K., & Schlücker, S. (2011). Hydrophilically stabilized gold nanostars as SERS labels for tissue imaging of the tumor suppressor p63 by immuno-SERS microscopy. *Chemical Communications,* 47(14), 4216-4218. http://www.ncbi.nlm.nih.gov/pubmed/21359379 (Last access date: June 2013). http://dx.doi.org/10.1039/c0cc05229a

Shanov, V., Yun, Y.H., & Schulz, M.J. (2006). Synthesis and characterization of carbon nanotube materials (review). *Chemical Vapor Deposition, 41(4),* 377-390.
http://adm1.uctm.edu/journal/j2006-4/01-Shanov-377-390.pdf

Shoshi, A., Schotter, J., Schroeder, P., Milnera, M., Ertl, P., Charwat, V., et al. (2012). Biosensors and Bioelectronics Magnetoresistive-based real-time cell phagocytosis monitoring. *Biosensors and Bioelectronics, 36,* 116-122. http://dx.doi.org/10.1016/j.bios.2012.04.002

Staleva, H., Skrabalak, S.E., Carey, C.R., Kosel, T., Xia, Y., Hartland, G.V., et al. (2009). Nanophotonics: plasmonics and metal nanoparticles. *Physical Chemistry Chemical Physics (PCCP), 11(28),* 5866. http://www.ncbi.nlm.nih.gov/pubmed/19588004 (Last access date: May 2013). http://dx.doi.org/10.1039/b911746f

Sudibya, H.G., Ma, J., Dong, X., Ng, S., Li, L.J., Liu, X.W., et al. (2009). Interfacing glycosylated carbon-nanotube-network devices with living cells to detect dynamic secretion of biomolecules. *Angewandte Chemie, 48(15),* 2723-2726. http://www.ncbi.nlm.nih.gov/pubmed/19263455 (Last access date: June 2013).

Thévenot, D.R., Toth, K., Durst, R.A., & Wilson, G.S. (1999). Electrochemical Biosensors: Recommended Definitions and Classification (Technical Report). *Pure and Applied Chemistry, 71(12),* 2333-2348. http://dx.doi.org/10.1351/pac199971122333

Titz, È., Matz, H., Drung, D., Hartwig, S., Groû, H., & R Ko, R. (1999). A squid measurement system for immunoassays. *Applied Superconductivity, 6(99),* 577-583.

Tripp, R.A., Dluhy, R.A., & Zhao, Y. (2008). Novel nanostructures for SERS biosensing recent progress in SERS biosensing is given in this article. *Review Literature And Arts Of The Americas, 3(3),* 31-37.

Tumanski, S. (2007). Induction coil sensors-a review. *Measurement Science and Technology, 18(3),* R31-R46.
http://stacks.iop.org/0957-0233/18/i=3/a=R01?key=crossref.46d204a2215e615d8cf1ed4b25518f39
http://dx.doi.org/10.1088/0957-0233/18/3/R01

Vohs, J.K., & Fahlman, B.D. (2007). Advances in the controlled growth of nanoclusters using a dendritic architecture. *New Journal of Chemistry, 31(7),* 1041.
http://xlink.rsc.org/?DOI=b616472m http://dx.doi.org/10.1039/b616472m

Volmer, M., & Avram, M. (2013). Signal dependence on magnetic nanoparticles position over a planar Hall effect biosensor. *Microelectronic Engineering, 108,* 116-120.
http://linkinghub.elsevier.com/retrieve/pii/S0167931713001664 (Last access date: January 2014). http://dx.doi.org/10.1016/j.mee.2013.02.055

Wang, J., Sui, M., & Fan, W. (2010). Nanoparticles for tumor targeted therapies and their pharmacokinetics. *Current Drug Metabolism, 11(2),* 129-141.
http://www.ncbi.nlm.nih.gov/pubmed/20359289 http://dx.doi.org/10.2174/138920010791110827

Wegner, K.D., Lanh, P.T., Jennings, T., Oh, E., Jain, V., Fairclough, S.M., et al. (2013). Influence of Luminescence Quantum Yield, Surface Coating, and Functionalization of Quantum Dots on the Sensitivity of Time-Resolved FRET Bioassays. *ACS Applied Materials & Interfaces, 5(8),* 2881-2892. http://www.ncbi.nlm.nih.gov/pubmed/23496235 http://dx.doi.org/10.1021/am3030728

West, J.L., & Halas, N.J. (2003). Engineered nanomaterials for biophotonics applications: improving sensing, imaging, and therapeutics. *Annual Review of Biomedical Engineering, 5,* 285-292. http://www.ncbi.nlm.nih.gov/pubmed/14527314 (Last access date: June 2013).

Winnik, F.M., & Maysinger,D. (2013). Quantum dot cytotoxicity and ways to reduce it. *Accounts of chemical research, 46(3),* 672–80. http://www.ncbi.nlm.nih.gov/pubmed/22775328 (Last access date: January 22, 2014).

Xia, Y., & Lim, B. (2010). Nanotechnology: Beyond the confines of templates. *Nature, 467(7318),* 923-924. http://dx.doi.org/10.1038/467923a

Xu, L., Yu, H., Han, S.J., Osterfeld, S., White, R.L., Pourmand, N., & Wang, S.X. (2008). Giant Magnetoresistive Sensors for DNA Microarray. *IEEE Transactions on Magnetics, 44(11),* 3989-3991.
http://www.pubmedcentral.nih.gov/articlerender.fcgi?artid=2933090&tool=pmcentrez&rendertype=abstract
http://dx.doi.org/10.1109/TMAG.2008.2002795

Yang, H.W., Hua, M.Y., Chen. S.L., & Tsai, R.Y. (2013). Reusable sensor based on high magnetization carboxyl-modified graphene oxide with intrinsic hydrogen peroxide catalytic activity for hydrogen peroxide and glucose detection. *Biosensors & Bioelectronics, 41,* 172-179. http://www.ncbi.nlm.nih.gov/pubmed/22959012 (Last access date: June 2013). http://dx.doi.org/10.1016/j.bios.2012.08.008

Yang, W., Thordarson, P., Gooding, J.J., Ringer, S.P., & Braet, F. (2010). Carbon nanotubes for biological and biomedical applications. *Nanotechnology, 18(41),* 412001. http://dx.doi.org/10.1088/0957-4484/18/41/412001

Yang, Z., Tjiu, W.W., Fan, W., & Liu, T. (2013). Electrodepositing Ag nanodendrites on layered double hydroxides modified glassy carbon electrode: Novel hierarchical structure for hydrogen peroxide detection. *Electrochimica Acta, 90,* 400-407. http://linkinghub.elsevier.com/retrieve/pii/S0013468612019974 (Last access date: June 2013). http://dx.doi.org/10.1016/j.electacta.2012.12.038

Yantzi, J.D., & Yeow, J.T.W. (2005). Carbon nanotube enhanced pulsed electric field electroporation for biomedical applications. *2005 IEEE International Conference Mechatronics and Automation, 4,* 1872-1877. http://ieeexplore.ieee.org/lpdocs/epic03/wrapper.htm?arnumber=1626847 http://dx.doi.org/10.1109/ICMA.2005.1626847

Yi, Z., Chen, S., Chen, Y., Luo, J., Wu, W., Yi, Y., et al. (2012). Preparation of dendritic Ag/Au bimetallic nanostructures and their application in surface-enhanced Raman scattering. *Thin Solid Films, 520(7),* 2701-2707. http://linkinghub.elsevier.com/retrieve/pii/S0040609011020554 (Last access date: June 2013). http://dx.doi.org/10.1016/j.tsf.2011.11.042

Yu, M.K., Jeong, Y.Y., Park, J., Park. S., Kim, J.W., Min, J.J., et al. (2008). Drug-loaded superparamagnetic iron oxide nanoparticles for combined cancer imaging and therapy in vivo. *Angewandte Chemie, 47(29),* 5362-5365. http://www.ncbi.nlm.nih.gov/pubmed/18551493 (Last access date: March 2012).

Yu, Y., Chen, X., Wei, Y., Liu, J.H., Yu, S.H., & Huang, X.J. (2012). CdSe Quantum Dots Enhance Electrical and Electrochemical Signals of Nanogap Devices for Bioanalysis. *Small Weinheim an der Bergstrasse Germany, 8(21),* 3274-3281. http://www.ncbi.nlm.nih.gov/pubmed/22761032 http://dx.doi.org/10.1002/smll.201200487

Zelada-Guillén, G.A., Riu, J., Düzgün, A., & Rius, F.X. (2009). Immediate detection of living bacteria at ultralow concentrations using a carbon nanotube based potentiometric aptasensor. *Angewandte Chemie, 48(40),* 7334-7337. http://www.ncbi.nlm.nih.gov/pubmed/19569156 (Last access date: May 2013).

Zhang, D., Carr, D.J., & Alocilja, E.C. (2009). Fluorescent bio-barcode DNA assay for the detection of Salmonella enterica serovar Enteritidis. *Biosensors and Bioelectronics, 24(5),* 1377-1381. http://www.ncbi.nlm.nih.gov/pubmed/18835708 http://dx.doi.org/10.1016/j.bios.2008.07.081

Zhang, X., Guo, Q., & Cui, D. (2009). Recent advances in nanotechnology applied to biosensors. *Sensors (Basel, Switzerland), 9(2),* 1033-1053. http://www.pubmedcentral.nih.gov/articlerender.fcgi?artid=3280846&tool=pmcentrez&rendertype=abstract (Last access date: May 2013). http://dx.doi.org/10.3390/s90201033

Zhu, S.Q., Zhang, T., Guo, X.L., Wang, Q.L., Liu, X., & Zhang, X.Y. (2012). Gold nanoparticle thin films fabricated by electrophoretic deposition method for highly sensitive SERS application. *Nanoscale Research Letters, 7(1),* 613. http://www.pubmedcentral.nih.gov/articlerender.fcgi?artid=3502474&tool=pmcentrez&rendertype=abstract http://dx.doi.org/10.1186/1556-276X-7-613

Chapter 2

Bacterial membrane formation monitored with atomic force microscopy and quartz crystal microbalance

Nadja Krska, Jose L. Toca-Herrera[*]

Institute for Biophysics, Department for Nanobiotechnology, University of Natural Resources and Life Sciences Vienna (BOKU), Muthgasse 11, Vienna, A-1190, Austria.

nadja.krska@gmail.com, [*]Corresponding author: jose.toca-herrera@boku.ac.at

Doi: http://dx.doi.org/10.3926/oms.168

Referencing this chapter

Krska, N., Toca-Herrera, J.L. (2014). Bacterial membrane formation monitored with atomic force microscopy and quartz crystal microbalance. In M. Stoytcheva & J.F. Osma (Eds.). *Biosensors: Recent Advances and Mathematical Challenges*. Barcelona, España: OmniaScience, pp. 41-50.

1. Introduction

Proteins are the machinery of life (Alberts, Bray, Johnson, Lewis, Raff, Roberts et al., 1998). They have a prominent role in food, pharmaceutical, or cosmetic technology (Yampolskaya & Platikanov, 2006; McPherson, 2003). Protein structure relates to function. Thus, to be able to perform their biological function, proteins fold into a specific spatial conformation. However, a general problem that arises in the laboratory is that protein adsorption on surfaces can denature (Gray, 2004). Therefore protein adsorption (and protein kinetics) are useful experiments to gain insight over protein adsorption and functionality.

S-layers are composed of single (glycol)proteins and constitute the most common outermost envelope component in prokaryotic organisms. They show oblique, square and hexagonal lattice symmetry with pore size between 2-8 nm and a thickness of 5-25 nm (Sleytr, Messner, Pum & Sára, 1999). A very important property of isolated S-proteins in solution, which few proteins exhibit, is their ability to form regular protein crystals on different (soft) interfaces (i.e. silicon oxide, mica, polyelectrolytes, self-assembly monolayers, or lipid films (Delcea, Krastev, Gutberlet, Pum, Sleytr & Toca-Herrera, 2008; Delcea, Krastev, Gutberlet, Pum, Sleytr, Toca-Herrera et al., 2007; Toca-Herrera, Moreno-Flores, Friedmann, Pum & Sleytr, 2004; Gyŏrvary, Stein, Pum & Sleytr, 2003; Martín-Molina, Moreno-Flores, Perez, Pum, Sleytr & Toca-Herrera, 2006; Moreno-Flores, Kasry, Butt, Vavilala, Schmittel, Pum et al., 2008; Eleta-Lopez, Pum, Sleytr & Toca-Herrera, 2011). In the last years, the production of fusion proteins based on Sproteins have constituted an advance in sensor and nano technologies (Sleytr, Huber, Ilk, Pum Schuster & Egelseer, 2007; Kainz, Steiner, Sleytr, Pum & Toca-Herrera, 2010). In this work, we report on the adsorption and S-layer formation of the S-protein SbpA on silicon oxides substrates and at the air/water interface. Quartz Crystal Microbalance with Dissipation (QCM-D) monitoring was used to follow protein adsorption in real time, and estimate the adsorbed protein mass deposited per unit area. The nanostructure of the protein surface layers was investigated with atomic force microscopy measurements. Finally, the SbpA activity at the air water interface was determined with surface tension measurements.

2. Materials and methods

2.1. Materials

Silicon oxide coated quartz crystals (Q-sense, Gothenburg, Sweden) were used as substrates. SbpA (Mw=120 kDa), were isolated from Lysinibacillus sphaericus CCM 2177 according to a reported procedure (31). Protein recrystallization buffer was prepared with 5mM Trizma (Sigma) base and 10mM CaCl2 (98%, Sigma) and adjusted to pH=9 by titration.

2% sodium dodecyl sulphate, SDS, (99%, Fluka) and Hellmanex II (2%, Hellma) were used as cleaning solutions. Aqueous solution of 100mM NaCl (Sigma) was used as media in AFM experiments.

Silicon oxide treatment. Silicon substrates with native silicon oxide layers were cleaned in 2% sodium SDS for 30 minutes, rinsed with ultrapure water (Barnstead) and dried under a stream of nitrogen. Afterwards the substrates were treated with ultraviolet radiation (Bioforce Nanoscieces) for other 30 minutes before silane modification or protein self-assembling.

S-protein adsorption. The S-protein solution was isolated as explained in reference (Sleytr, Sara, Küpcü & Messner, 1986). Due to the ability of S-proteins to self-assemble in solution, the protein extract solution (1mg/ml) was centrifuged at 5 rpm for 5 minutes to separate S-protein monomers from self-assembly products. The protein concentrate was determined with a Nanodrop spectrometer (Thermo Scientific, Wilmintong, USA). Before starting the experiments, the supernatant was diluted using the appropriate amount of recrystallizing buffer. On one hand, in-situ QCM-D experiments were carried out by protein solution injection into the experimental set up, once the substrates were stabilized in tris-buffer. On the other hand, ex-situ experiments were done using substrates where S-protein had been previously recrystallized. Small volume humidity chambers were used for that purpose preventing water evaporation. The protein was incubated between one and three hours at room temperature and afterwards the substrates were rinsed with recrystallizing buffer in order to remove excess of protein.

2.2. Apparatus

2.2.1. Quartz Crystal Microbalance with Dissipation monitoring (QCM-D)

The QCM-D set up consists of a thin AT-cut piezoelectric quartz crystal sandwiched between two electrodes. The crystal oscillates in shear modus when AC voltage is applied. The interesting feature is that the resonant frequency of the sensor depends on the total mass, which includes the entrapped water. For many cases, the measured change in frequency is proportional to the adsorbed mass (more details are given in the results section). This technique also permits to evaluate the dissipated frictional energy delivering information about the viscoelastic properties of the absorbed molecules on the quartz crystal (more details are given in the results section). In our investigation, QCM-D (Q-sense E4, Gothenburg, Sweden) was used to carry out real time/in situ experiments.

2.2.2. Atomic Force Microscopy (AFM)

The sensor of the AFM consists on a sharp tip mounted on a soft cantilever of a specific spring constant. In particular, due to the interaction forces between the tip and the sample it is possible to obtain topographical information at the nanoscale and quantify interaction forces down to dozens of pN. The optical detection system of the AFM consists of a photodiode that follows the cantilever deflection (through the reflection of the laser beam at its back side) while scanning the sample. The photodiode electrical signal is later processed by a computer that keeps the force on the tip constant by generating a feedback signal for the piezoscanner.

The original images presented in this manuscript were recorded in contact (at different scan rates) in 100mM NaCl aqueous solution, at room temperature using a Nanoscope V controlled Multimode AFM (Bruker, Santa Barbara, USA). Back side gold coated silicon nitride (Si3N4, NP-S, Bruker) cantilevers with nominal spring constant of 0.06 N/m were used.

3. Results and discussion

SbpA adsorption was monitored in real time by QCM-D. Figure 1 illustrates the variation of the frequency (a) and dissipation (b) as a function of time for three protein concentrations. When the protein solution is injected in the experimental chamber, the frequency decreases due to protein deposition, while the dissipation increases. Once no further changes in frequency and dissipation are observed, the excess of protein is removed from the chamber by rinsing with fresh buffer solution. This process hardly affects frequency and dissipation meaning that the freshly formed protein layer does remain on the silicon dioxide surface.

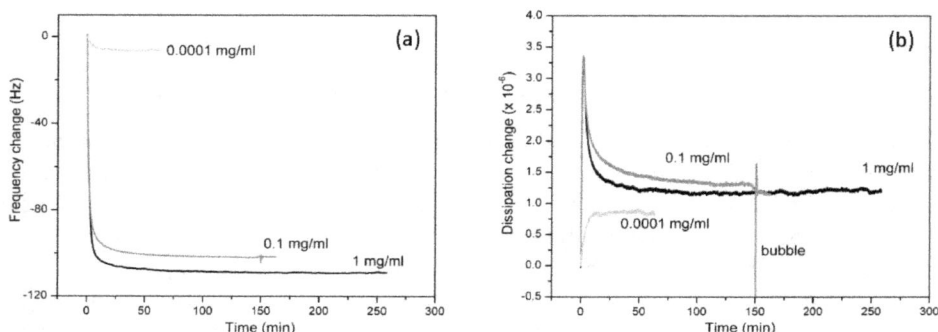

Figure 1. QCM-D measurement. (a) frequency change for SbpA adsorption on silicon dioxide as a function of time. A frequency shift of about 100 Hz for the two largest concentration is observed. Note the low frequency value obtained for 0.0001 mg/ml. (b) dissipation change for three SbpA concentrations as a function of time. The largest concentrations have a similar final value after protein layer is formation. (The straight line refers to an air bubble that appeared when rinsing with buffer)

In order to quantify the incorporated mass one should make certain assumptions: i) the protein layer is homogeneously distributed and rigid enough to avoid any oscillatory deformation, and ii) the added protein mass is lower than the quartz crystal itself.

Furthermore, if the overall dissipation-frequency shift ratio is smaller than 0.2 x 10^{-6}Hz^{-1}, (Gläsmäster, Larsson, Höök & Kasemo, 2002) the Sauerbrey equation can be used to determine adsorbed mass once the frequency change is known:

$$\Delta m = -\frac{C \cdot \Delta F}{n}$$

here Δm is the mass surface density (ng/cm^2), C is a constant that depends on the intrinsic properties of the sensor (-17.7 ng/(Hz·cm^2)), ΔF is the frequency change and n is the overtone number. In this way the mass per unit area calculated using the 5th overtone (ΔF), is 1930 ng/cm^2, 1800 ng/cm^2, and 108 ng/cm^2 for 1 mg/ml, 0.1 mg/ml and 0.0001 mg/ml respectively. (Note that the smallest concentration did not lead to a compact protein layer, as the force-distance curves will show later, and therefore the values reported are just illustrative.).

Figure 2 shows the variation of the adsorbed mass with time for SbpA adsorption on silicon dioxide.

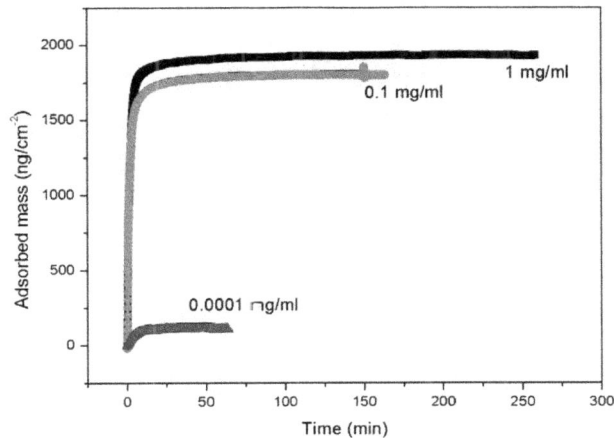

Figure 2. SbpA mass adsorption on silicon dioxide as a function of time for three concentration. The adsorption is constant after ca. 60 minutes. The difference in mass for the two largest concentrations is about 100 ng/cm^2

A quick analysis of the exponential growing shows that the adsorption for 1 mg/ml is the fastest one. Similar values for the adsorbed mass (and also frequency and dissipation) were obtained for SbpA adsorption studies on (poly(sodium 4-styrenesulfonate)) and self-assembled monolayers (Delcea et al., 2008; Eleta-Lopez et al., 2011).

The (surface) structure of the adsorbed protein on silicon dioxide QCM-D sensors was determined with atomic force microscopy. Figure 3 shows AFM deflection images of the three studied protein concentrations. On the left, it can be seen that the smallest protein concentration (0.0001 mg/ml) does not form a crystalline layer. The image of the middle corresponds to a protein concentration of 0.1 mg/ml. Although the image shows bumpy S-layer areas, the characteristic P4 pattern of the Slayer is to be recognised. On the right, the highest protein concentration (1 mg/ml) leads to a more homogeneous protein layer of clear P4 structure similar to the crystalline structure found in bacteria (Sleytr, Sára, Pum, & Schuster, 2001).

Figure 3. AFM deflection images measured in contact mode of adsorbed SbpA on silicon dioxide. All the measurements were performed in 100 mM NaCl aqueous solution. The result of the lowest concentration (0.0001 mg/ml), left image, is an indication that the adsorbed proteins on the QCM-D crystal do not form a crystalline protein layer. On the contrary, S-layer formation took place for higher concentrations (0.1 mg/ml and 1 mg/ml), middle and right images respectively

Recrystallization processes normally occur in two stages, nucleation and growth (McPherson, Kuznetsov, Malkin & Plomp, 2003). While the substrate-protein interaction is crucial to the formation of the nucleation points, the protein-protein interaction might be more important for the incorporation of new proteins and coalescence of 8 crystal domains leading to the final compacted protein layer. QCM-D frequency-dissipation curves is an easy and qualitative way to link the layer formation and the mechanical properties of the layer. Figure 4 illustrates the variation of the dissipation as function of the frequency change (adsorption of mass). Few proteins adsorbed for a concentration of 0.0001 mg/ml leading to a small increase in the dissipation. More interesting is the tendency for highest concentrations. The dissipation starts to rise at the same moment that proteins adsorb on the substrate in a very similar way until reaching a maximum value of about 3.5×10^{-6}. In this first part, protein random adsorption with high degrees of freedom of molecular movement and a hydrodynamic process (depending on the protein concentration) could be the reason for the observed effect. After the maximum of the curve, the dissipation decreases while more protein is being adsorbed on the substrate (the frequency becomes more negative). This can be interpret as the start of formation of a "rigid" structure (of less molecular motion freedom). Finally, at frequency values of about 90 and 100 Hz the dissipation reaches its lower value, which means that the S-layer is formed (as shown by AFM measurements).

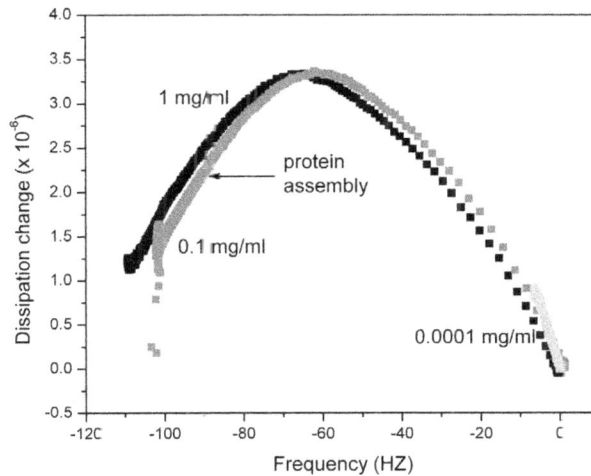

Figure 4. Dissipation change versus frequency shift for SbpA adsorption on silicon dioxide. The dissipation shows a maximum for a frequency value of about 60 Hz, showing the possible moment at which protein self-assembly takes place. Note that a dissipation value of zero would mean that the protein layer has similar mechanical properties as the QCM-D substrate

In order to get more insight about the mechanical properties of the adsorbed protein layer, AFM force distances curves were carried out (see Figure 5). Force-distance curves at a loading rate of 1000 nm/s showed that the protein layer is stable for the two highest proteins while soft and elastic protein domains were detected for protein concentration of 0.0001 mg/ml. On one hand, the left image of figure 5 confirms that the AFM tip senses a hard wall (in this case the S-layer) and that practically no energy is lost in the process (the approaching and the retracting curves overlap each other, blue and red respectively). On the other hand, on the right image, the red curve (when the AFM tip leaves the surface) shows a saw-tooth pattern typical for polymer stretching (as well as adhes on energy) with forces of about 100 pN and extension lengths of tenths of nm. This curve also shows a larger curvature before the contact zone between the AFM tip and the substrate indicating a higher repulsive force (probably of electrostatic and entropic nature) and a different "protein layer" of softer texture.

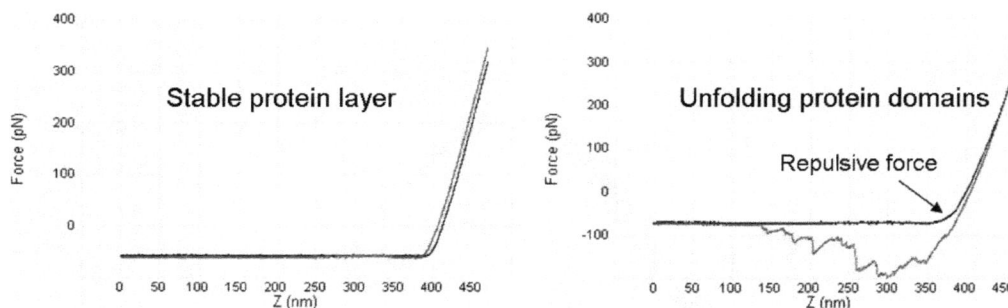

Figure 5. On the left, a force-distance curve showing the robustness of a protein crystal (S-layer). The approaching curve (blue) and the retracting curve (red) practically overlap each other showing no hysteresis or elastic domains. On the right, it can be seen that AFM tip could not easily be removed from the surface. I this case the is pulling proteins out of the substrate after applying a load of 300 pN. Note that the adhesion/unfolding forces are about 100 pN

4. Conclusions

S-protein adsorption on silicon dioxide and at the air-water interface has been studied with atomic force microscopy and quartz crystal microbalance and tensiometry. A first analysis of the results indicate that protein adsorption is primary governed by diffusion (until saturation is reached at the highest concentration). Low SbpA protein concentrations (0.0001 mg/ml) do not form a crystalline protein layer but a soft one of elastic nature as AFM image and force-curves show. On the contrary, protein concentrations of 0.1 mg/ml and 1 mg/ml form a "rigid" crystalline layer corresponding to a surface mass density of about 1800 ng/cm^2 and 1930 ng/cm^2 respectively. Finally, it has been confirmed that SbpA adsorbs at the air-water interface reducing in about 20 mN/m the surface tension of water for the highest protein concentration. New experiments would be needed to refine the protein adsorption on silicon surfaces and to map a molecular model of protein adsorption at such interfaces.

Acknowledgments

NK as Talente Grant holder thanks die Österreichische Forschungsförderungsgesellschaft (FFG) for financial support.

References

Alberts, B., Bray, D., Johnson, A., Lewis J., Raff, M., Roberts, K., et al. (1998). *Essential Cell Biology: an introduction to the Molecular Biology to the Cell.* Garland Publishing Inc New York & London. 133-181.

Delcea, M., Krastev, R., Gutberlet, T., Pum, D., Sleytr, U.B., & Toca-Herrera, J.L. (2008). Thermal stability, mechanical properties and water content of bacterial protein layers recrystallized on polyelectrolyte multilayers. *Soft Matter, 4,* 1414-1421. http://dx.doi.org/10.1039/b719408k

Delcea, M., Krastev, R., Gutberlet, T., Pum, D., Sleytr, U.B., Toca-Herrera, J.L., et al. (2007). Mapping bacterial surface layers affinity to polyelectrolytes through the building of hybrid macromolecular structures. *Nanotechnol., 7,* 4260-4266.

Eleta-Lopez, A., Pum, D., Sleytr. U.B., & Toca-Herrera, J.L. (2011). Influence of surface chemistry and protein concentration on adsorption rate and S-layer crystal formation. *Phys. Chem. Chem. Phys., 13,* 11905-11913. http://dx.doi.org/10.1039/c1cp00052g

Gläsmäster, K., Larsson, Ch., Höök, F., & Kasemo, B. (2002). Protein adsorption on supported phospholipid bilayers. *J. Colloid Interface Sci., 246,* 40-47. http://dx.doi.org/10.1006/jcis.2001.8060

Gray, J.J. (2004).The interaction of proteins with solid interfaces. *Curr. Opin. Struct. Biol., 14,* 100-115. http://dx.doi.org/10.1016/j.sbi.2003.12.001

Györvary, E.S., Stein, O., Pum, D., & Sleytr, U.B. (2003). Self-assembly and recrystallization of bacterial S-layer proteins at silicon supports imaged in real time by atomic force microscopy. *J. Microsc., 212,* 300-306. http://dx.doi.org/10.1111/j.1365-2818.2003.01270.x

Kairz, B., Steiner, K., Sleytr, U.B., Pum, D., & Toca-Herrera, J.L. (2010). Fluorescent S-layer protein colloids. *Soft Matter, 6,* 3809-3814. http://dx.doi.org/10.1039/c0sm00008f

Martín-Molina, A., Moreno-Flores, S., Perez, E., Pum, D., Sleytr, U.B., & Toca-Herrera, J.L., (2006). Structure, surface interactions and compressibility of bacterial S-layers through surface forces techniques. *Biophys. J., 90,* 1821-1829. http://dx.doi.org/10.1529/biophysj.105.067041

McPherson, A. (2003). Macromolecular crystallization in the structural genomics era. *J. Struct Biol., 142,* 1-2. http://dx.doi.org/10.1016/S1047-8477(03)00033-9

McPherson, A., Kuznetsov, Y.G., Malkin, A., & Plomp, M. (2003). Macromolecular crystal growth as revealed by atomic force microscopy. *J. Struct. Biol., 142,* 32-46. http://dx.doi.org/10.1016/S1047-8477(03)00036-4

Moreno-Flores, S., Kasry, A. Butt, H.J., Vavilala, C., Schmittel, M., Pum, D., et al. (2008). From native to non-native two-dimensional protein lattices via underlying hydrophilic/hydrophobic nanoprotrusions. *Angew. Chem. Int. Ed., 47,* 4707-1717. http://dx.doi.org/10.1002/anie.200800151

Sleytr, U.B., Huber, C., Ilk, N, Pum, D. Schuster, B., & Egelseer, E.M. (2007). S-layers as a tool kit for nanobiotechnological applications. *FEMS Microbiol. Let., 267,* 131-144. http://dx.doi.org/10.1111/j.1574-6968.2006.00573.x

Sleytr, U.B., Messner, P., Pum, D., & Sára, M. (1999). Crystalline Bacterial Cell Surface Layers (S Layers): From Supramolecular Cell Structure to Biomimetics and Nanotechnology. *Angew. Chem. Int. Ed., 38,* 1034-1054. http://dx.doi.org/10.1002/(SICI)1521-3773(19990419)38:8<1034::AID-ANIE1034>3.0.CO;2-#

Sleytr, U.B., Sara, M., Küpcü, S., & Messner, P. (1986). Structural and chemical characterization of S-layers of selected strains of Bacillus stearothermophilus and Desulfotomaculum nigrificans. *Arch. Microbiol., 146,* 19-24. http://dx.doi.org/10.1007/BF00690152

Sleytr, U.B., Sára, M., Pum, D., & Schuster, B. (2001). Characterization and use of crystalline bacterial cell surface layers. *Prog. Surf. Sci., 68,* 231-278. http://dx.doi.org/10.1016/S0079-6816(01)00008-9

Toca-Herrera, J.L., Moreno-Flores, S., Friedmann, J., Pum, D., & Sleytr, U.B. (2004). Chemical and Thermal Denaturation of Crystalline Bacterial S-Layer Proteins. *Microsc. Res. Techniq., 65,* 226-234.

Yampolskaya, G., & Platikanov, D. (2006). Proteins at Fluid Interfaces: Adsorption Layers and Thin Liquid Films. *Adv. Colloid Interf. Sci, 128-130,* 159-183. http://dx.doi.org/10.1016/j.cis.2006.11.018

Chapter 3

Whole cell biosensors

César A. Hernández, Johann F. Osma

CMUA. Department of Electrical and Electronics Engineering. Universidad de los Andes, Colombia.

ca.hernandez11@uniandes.edu.co

Doi: http://dx.doi.org/10.3926/oms.192

Referencing this chapter

Hernández C.A., Osma, J.F. (2014). Whole cell biosensors. In M. Stoytcheva & J.F. Osma (Eds.). *Biosensors: Recent Advances and Mathematical Challenges*. Barcelona: OmniaScience, pp. 51-96.

1. Introduction

The term biosensors can be used in a very broad sense for a large set of devices that use biologically active components, conjugated with appropriate transducers, which bear the detection of sought substances (Thévenot, Toth, Durs & Wilson, 2001). A recurrent example when referring to biosensors is the coal-mines canary, often used to depict the early use of living species to provide environmental information of potential hazards. Such example has been expedient to spot the advantages of higher organisms on the detection of menaces, reports (Van der Schalie, Gardner, Bantle, De Rosa, Finch, Reif et al., 1999; Rabinowitz, Gordon, Chudnov, Wilcox, Odofin, Liu et al., 2006; Gubernot, Boyer & Moses, 2008; Rabinowitz, Scotch & Conti, 2009) render an extensive guide on case-study and case of application on animals that can be used as biological markers for early warning systems, in Table 1, a compiled list of examples using animals as sentinels is provided. Physiological and behavioral responses as exhibited by aquatic organisms are a benchmark of this postulate (Van der Schalie, 1977; Morgan & Young, 1984; Kramer, 2009): Architectures using fishes (Morgan & Eagleson, 1983), mussels (Butterworth, Gunatilaka & Gonsebatt, 2001) and invertebrates (Lechelt, Blohm, Kirschneit, Pfeiffer, & Gresens, 2000) are described as reliable markers for commercially available systems (Fish Toximeter- Detection of Toxic Substances in the Water - Bbe Moldaenke, 2013; Musselmonitor (Mosselmonitor): a Biological Early Warning System, 2013).

Sentinel Incidents					References[†]
Species	**Toxicant**	**Country**	**Date**	**Related events**	
Canaries	Carbon monoxide	England	1870s		(Burrell & Seibert; 1916; Schwabe, 1984)
Cattle	Smog	England			(Veterinarian, 1874a, 1874b)
Cattle	Fluoride	England	1910s		
Horses	Lead	United States			
Cattle	TCE	Scotland			(Haring & Meyer, 1915; Stockman, 1916; Holm, Wheat, Rhode & Firch, 1953; Medtronics Associates, 1970)
Cats	Mercury	Japan	1950s		(Kurland, Faro & Siedler, 1960)
Birds	DDT	United States			
Cattle	Smog	England			
Chickens	PCBs	Japan	1960s	Silent Spring published (1962)	(Carson, 1962; Kuratsune, Yoshimura, Matsuzaka &. Yamaguchi, 1972)
Sheep	OP agents	United States			(Van Kampen, James, Rasmussen, Huffaker & Fawcett, 1969)

Sentinel Incidents					References[†]
Species	Toxicant	Country	Date	Related events	
Horses and other animals	Dioxin	United States	1970s	NCR Symposium on Pathobiology of Environmental Pollutants: Animal Models and Wildlife Monitors (1979)	(Case & Coffman, 1973; Carter, Kimbrough, Liddle, Cline, Zack, Barthel et al., 1975; Northeastern Research for Wildlife Diseases, Registry of Comparative Pathology & Institute of Laboratory Animal Resources (U.S.), 1979)
Dairy cattle	PBBs	United States			(Jackson & Halbert, 1974; Welborn, Allen, Byker, DeGrow, Hertel, Noordhoek et al.,1975)
Sheep	Zinc	Peru	1980s	Task Force on Environmental Cancer and Heart and Lung Disease established (1981) NAS risk assessment paradigm (1983) NCI report, Use of Small Fish Species in Carcinogenicity Testing (1984)	National Research Council, 1983)
Aligators	DDT, dicofol	United States	1990s	NRC report, Animals as Sentinels of Environmental Health Hazards (1991)	(National Research Council, 1991; U.S. Congress 1993; Jr & Gross, 1994)
Fish	*Pfiesteria* toxins	United States		Public Law 103-43 enacted (1993) NIEHS establishes ad hoc ICCVAM (1993) ICCVAM report, Validation and Regulatory Acceptance of Toxicological Methods (1997)	(Interagency Coordinating Committee on the validation of alternative Methods, 1997)

Abbreviations: TCE, tetrachloroethylene; PCBs, polychlorinated biphenyls; OP, organophosphate; NCR, National Research Council; PBBs, polybrominated biphenyls; NAS, National Academy of Sciences; NCI, National Cancer Institute; NIEHS, National Institute of Environmental Sciences; ICCVAM, Interagency Coordinating Committee on the Validation of Alternative Methods
† All references as cited by (Van der Schalie et al., 1999)

Table 1. Timeline of examples of animals as sentinels of environmental toxicants and noteworthy events, extracted from (Van der Schalie et al., 1999)

The thermodynamic reaction path provides an advantageous guideline to allocate two well defined strategies, namely: Stimuli-Response-Based (SRB) and Biotransformation-Based (BtB) strategies, which respond to the equation posed by Willard Gibbs for describing the free energy of a reaction (Gibbs, 1873). Additionally, the proposed strategies acknowledge the signal reading as direct (SRB) and indirect (BtB) transduction methods. Some advantages and disadvantages of using both strategies are listed in Table 2.

	SRB Strategy	**BtB Strategy**
Advantages	• Direct assessment of electrical measures • Faster response • Less complex measurement mechanism • No need for additional reagents/elements for measuring • Lower probability for occurrence of side-effect reactions • Swifter usage for the assembled device	• Yield a ready amplified signal • No need for immobilization of cells • Wide variety of response signals • Some cases don't need to use measurement equipment
Disadvantages	• Low level signals • High noise to signal ratio (NSR) • Need for an output signal external amplifier for portable applications • Need for immobilization and precise biomass control • Need for measuring devices to read the output signal	• Indirect measurement • The output signal is mediated by complementary reactions • Response dependent on the reaction and metabolic pathway time • Usual need for additional reagents/elements for measuring • Higher probability for occurrence of side-effect reactions

Table 2. Advantages and disadvantages of the SRB and BtB strategies

Some of the more relevant issues of tissue-based biosensors are described in (Arnold & Rechnitz, 1980; Rechnitz & Ho, 1990; Wijesuriya & Rechnitz, 1993; Safronova, Khichenko & Shtark, 1995; Rudolph & Reasor, 2001), a classification of mammalian cells is presented in Table 3 according to their function and mechanism of biosensing, they are set as an example of the multiple nature of tissue used for biosensing purposes, multiple mechanisms are used in order to harvest information derived from tissues, namely transduction methods, ranging from measurements that rely on electric, physiological, metabolic, optical and genetic changes (Arnold, 1986; Hansen, Wittekindt & Sherry, 2009; Acha, Andrews, Huang, Sardar & Hornsby, 2010; Belkin & Gu, 2010).

A ready-witted approach to biosensing, has been made by the isolation and utilization of biocatalysts to attain a variety of reactions that allow the addressing of different substrates (Milner & Maguire, 2012), which pose particular relevance for health, environment and industrial purposes. Among the used biocatalysts in biosensors, enzymes, which expedite and manage chemical reactions, to ensure the bearing and survival of whole-cells, have drawn the attention of academic and industrial communities mainly due to their high specificity, portability, miniaturization capacity, ease of in-situ utilization and fast response (Wilson & Hu, 2000; Newman & Setford, 2006).

The question that arises is then: Why is it appealing to deepen into whole-cell biosensors? Consider the whole-cell as a natural factory of biocatalysts: usually, the method for acquiring isolated biocatalysts, like the aforesaid enzymes, requires a process of separation and purification from the raw whole-cell strain or tissue, which is time and resource consuming. Furthermore, the ensuing enzyme sometimes requires cofactors and coenzymes, to carry out a complete reaction or to recognize a substrate; such are paired by adding them to the purified enzyme or by combining a supplementary enzyme, which implies a recycling process to attain the required cofactor, then eliciting the need for further separation and purification steps.

Function/ mechanism at cellular levels	Cell types	Primary signals derived from cells	Device/method used for secondary transduction of signals	Application areas	Strategy	References
Excitable/ electrogenic	Neuron, cardiac cells, neuronal network	pH, flow of ions	LAPS, microelectrodes	Drug discovery and testing (dose response) Toxicology Pharmacology	SRB	(Pancrazio, Gray, Shubin, Kulagina, Cuttinom, Shaffer et al., 2003; Liu, Cai, Xu, Xiao, Yang & Wang, 2007; Parviz & Gross, 2007)
Electrical responses	Epithelial cells, cardiomyocytes, neurons	Electric current, flow of ions	ECIS, Impedance, IDES, electrophysiology and electric potential, BERA	Bio-assays Drug discovery and testing (dose response) Toxicology	SRB	(Giaever & Keese, 1993; Gilchrist, Giovangrandi, Whittington & Kovacs, 2005; Asphahani & Zhang, 2007; Kloss, Fischer, Rothermel, Simon & Robitzki, 2008; Slaughter & Hobson, 2009)
Cellular receptors	Epithelial cells, hepatocytes, stem cells, mast cells, mononuclear cells, T- or B-cells	pH, alteration of molecules within cells	Cell-signaling molecules, LAPS, optical methods, DPSCA	Pathogen and toxin testing Combinatorial chemistry Bio-assays Toxicology Environmental monitoring Biosecurity	SRB/ BtB	(Kamei, Haruyama, Mie, Yanagida, Aizawa & Kobatake, 2003; Rider, Petrovick, Nargi, Harper, Schwoebel, Mathews et al. 2003; Trask, Baker, Williams, Nickischer, Kandasamy, Laethem et al., 2006; Curtis, Naal, Batt, Tabb & Holowka, 2008)
Cellular metabolism	Epithelial cells, hepatocytes, stem cells	pH, ion channel, molecular flux	pH-sensitive ISFETs, LAPS, ion-sensors	Drug discovery and testing (dose response) Toxicology Bio-assays	SRB/ BtB	(Xu, Ye, Qin, Xu, Li, Li et al., 2005; Ceriotti, Kob, Drechsler, Ponti, Thedinga, Colpo et al., 2007; Liu et al., 2007)
Cytotoxicity	Epithelial cells, endothelial cells, macrophages, myeloma, mononuclear cells, T- or B-cells	Changes in membrane integrity, cellular morphology	Optical methods, potentiometric methods	Pathogen and toxin testing Bio-assays Toxicology Environmental monitoring Biosecurity	BtB	(Banerjee, Lenz, Robinson, Rickus & Bhunia, 2008; Lee, Kumar, Sukumaran, Hogg, Clark & Dordick, 2008; Tong, Shi, Xiao, Liao, Zheng, Shen et al., 2009)
Genomic responses	Epithelial cells, endothelial cells, macrophages, myeloma cells, T- or B-cells	Changes in gene expression	Reporter gene assay, optical methods, cytometry	Bio-assays Drug discovery and testing (dose response) Toxicology Environmental monitoring	BtB	(Haruyama, 2006; Trask et al., 2006; East, Mauchline & Poole, 2008)

BERA, bioelectric recognition assay; DPSCA, double potential step chronoamperometry; ECIS, electric cell-substrate impedance sensing; IDES, interdigitated electrode structures; ISFET, ion-sensitive field effect transistor; LAPS, light-addressable potentiometric sensor

Table 3. Classification of cell-based biosensors based on the function and mechanism of action of the biosensing, modified from (Banerjee & Bhunia, 2009; Fleming, 2010)

Whole-cells contain a complete metabolic aggregate of enzymes, cofactors and coenzymes, constituting a well suited mechanism to assure chemical reactions that are fundamental for their function, in addition they self-regulate the recycling process for such substances; analog processes can be found in tissues, but the requirements related to maintenance and cost for culturing microorganisms are below from those of tissue cultures.

By following the route of different metabolic paths, where one or more enzymes are involved, whole-cells can yield a series of reactions that can be readily detected: One advantage on the usage of whole-cells is that very complex reactions can be attained by harnessing the presence of multiple enzymes in one single step. The resulting processes derived from the considered enzymatic activities, such as physiological responses – to mention motility, growth, respiration, digestion, among others – reveal prospects in which the advantageous properties of whole-cells are profited.

The selection of a relevant whole-cell for a biosensing application, primarily, would obey to the characteristics rendered by the selected strain on an observed environment, to set an example: Strains that grow on harsh environments yield to metabolic activities that require specific compounds to be performed and are copiously available on the targeted environment. The previous example is not a restriction, innate selectivity of whole-cells is not limited to strains that are harvested under extreme conditions, strains cultured in a friendlier environment can respond to very specific stimuli, such condition allows the screening of different strains as prospect candidates for a desired biosensing application.

In addition, whole-cells endure the modification and inclusion of regulatory mechanisms, such can be detached from other organisms or being synthetically tailored in a laboratory (Siegfried, 2011; Thomas, 2013). It is possible to alter genetic configuration of whole cells thus changing the enzymatic expression of a given strain, conveying the possibility to react upon different substrates or to include responses that can be easily monitored, the induced response is mediated for what is known as a bioreporter gene (Daunert, Barrett, Feliciano, Shetty, Shrestha & Smith-Spencer, 2000; Leveau & Lindow, 2002; Belkin, 2003; Jansson, 2003; Sørensen, Burmølle & Hansen, 2006; Yagi, 2007; Salis, Tamsir & Voigt, 2009; van der Meer & Belkin, 2010).

Numerous reviews have been made to present the advances, methods and characteristics of whole-cell biosensors (D'Souza, 2001; Farré, Pasini, Carmen Alonso, Castillo & Barceló, 2001; Harms, Wells & van der Meer, 2006; Tecon & van der Meer, 2008; Reshetilov, Iliasov & Reshetilova, 2010; van der Meer, 2011; Su, Jia, Hou & Lei, 2011; Shimomura-shimizu & Karube, 2010); the following pages present a general approach to the fundamentals and applications of whole-cell biosensors, the first section refers to the principles on whole-cell sensing, it deepens into the concept of enzymatic catalysis induced in living cells and divides the sensing strategies in two: Stimuli-Response-Based whole-cell biosensors and Biotransformation-Based whole-cell biosensors. A similar division was proposed in (Aston & Turner, 1984), referred as direct and indirect systems, analogue to SRB and BtB strategies respectively, such division is not explained on the basis of thermodynamic properties, but the use of a secondary transduction method. This section is directed to the non-biology-related readers, and might be only a quick reference for biologists.

The consecutive section deals with the relationship in between whole-cell transduction and chemical transduction, benefiting from the proposed classification of the latter section, it serves as a bridge between disciplines, it is here explained the parameters to be aware of in both interactions. It settles the differences from electrochemical effects in whole-cells (amperometric, potentiometric, conductometric and impedance -sensors) and the physiological effects, either due to respirometry, external stimuli or bioreporters associated results.

The last section presents the current advances and the market-available options using whole cell biosensors, the chapter is closed with the discussion of the future treads and challenges that they offer to be developed.

2. General principles on whole-cell biosensors

Whole-cells possess the ability to conduct important changes on different substrates through very well defined succession of reactions, such transformations are profited as energy or as essential elements for vital processes of the cell. A single strain can interact with different substrates at the same time, each of which is guided on every instance by a very specific series of chemical reactions, which compose a metabolic pathway.

Enzymes are the mediators that promote the occurrence of such chemical reactions; they are set to complete a chain of activities that secure that the whole-cells accomplish metabolic and physiological responses for assuring survival. An enzyme-catalyzed reaction would be a naturally occurring event which presents a modified energetic route.

The enzyme production is concurrently regulated by the genetic code of the whole-cell, which is in ensemble a unique imprint for every different strain. The integrity of the genetic information is kept in what can be compared to a 'storage unit" i.e. the DNA, the information set itself is referred as the genome. While the genome can vary from subject to subject, the basic function structure remains the same, which ultimately will indicate the identity of a given strain. The sequential arrangement of the resulting aminoacid constitutes the enzymatic structure as a complete programming code (Siegfried, 2011).

The very specific 3D structures of enzymes, formed by the folding and coiling of aminoacids, form fragments that act as pockets disposed for the coupling of the substrate, such are known as active sites, they are endowed with unique geometric shape and an atomic disposition complementary to those of the substrate.

A general approach for substrate-enzyme binding mechanisms can be envisioned through the "lock-and-key theory" (Koshland, 1995): In 1894 Emil Fischer proposed that enzymes were rigid structures, they would act as a lock with a specific shape; the substrate, on the other hand, resembles a key: If a properly shaped active site is available then it will serve as an adequate key hole, thus, it would "unlock" the consequent reaction (Figure 1).

In 1958 David Koshland offered an alternative to the Fischer's postulate: He stated that enzymes are flexible rather than rigid, hence the active site is keenly changing its shape to adapt to the substrate (Figure 1), the reaction would only take place if there are certain number of chemical bonds constructed between the active site and the substrate and are aligned to the catalytic groups, substrates can bind through e.g. van der Waals forces, ionic-, hydrogen- and covalent bonds; these interactions are generally weak, but with many of these interactions taking place at the same time, it constructs a solid binding. Koshland's postulate is recognized as the "induced-fit theory" (Koshland, 1995).

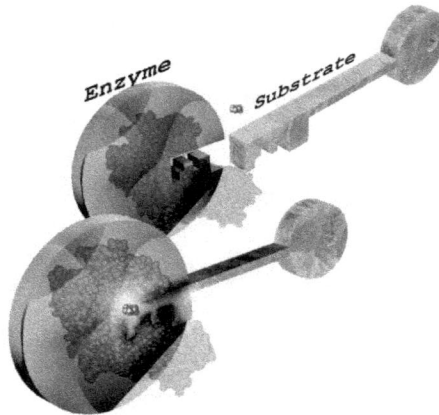

Figure 1. Representation of the lock and key theory: Only equally shaped substrates will bind to specific enzyme, circle represents the enzyme while the key represents the substrate. A comparison with the induced-fit is made, in blue, a 3D representation of an enzyme with flexible structure, the shape adapts to the substrate, depicted in red, though the active site. When the substrate is bonded the catalytic reaction takes place

The distinctive binding mechanism not only encloses the necessary conditions to understand selectivity, the core of the enzymatic process relies on the chemical reaction rate acceleration. As it is inferred from the induced-fit theory both mechanisms are dependent on the progression of molecular interactions with the substrate: While the initial enzyme-substrate interaction is feeble, the increasing amount of bindings formed between the active site and the substrate, which is only possible when an appropriate substrate is present, would prompt structural changes on it until the substrate is firmly attached to the enzyme, analogous mechanism can arrange multiple substrates, coenzymes and cofactors. The correct alignment might induce any of four acceleration mechanisms, to wit: Approximation of the reactants, covalent catalysis, general acid-base catalysis or the introduction of distortion or strain in the substrate (Jencks, 1987).

A reaction can be understood in terms of the thermodynamic properties of the system, relying on the first and second laws of thermodynamics, these are conservation of energy and increasing entropy: Willard Gibbs proposed in 1873 a model, analogous to the potential energy in classical mechanics, considering that a given system would have a thermodynamic potential to generate work under constant conditions of volume and temperature in a closed system, namely Gibbs free energy (denoted as G), originally denominated as available energy (Gibbs, 1873; Newman & Thomas-Alyea, 2012). The maximum amount of free energy derived in any chemical reaction is defined from the difference between the free energy of the products and the free energy of the reactants (ΔG), when the reaction consumes energy i.e. the free energy of the product is greater than the free energy of the reactants, it is said to be an endergonic reaction, that is, it requires an external source of energy e.g. heat. When the reaction releases energy it is said to be exergonic; it is thus considered a thermodynamically favorable reaction, meaning that it can spontaneously occur. Then, when an exergonic (spontaneous) reaction takes place, the entropy (S) will increase if no energy is provided to the system ($\Delta S > 0$). On the other hand, if no change in entropy is considered, the release of energy can only be attributed to the internal energy of the system, property known as enthalpy (H), as it is usually measured as heat, the difference between the initial and the resulting reaction state (ΔH) is known as the heat of a reaction. Accordingly, when the system's resulting process releases energy, it is known as exothermic (ΔH<0), on the contrary case it is called endothermic (ΔH>0); it is important to stress that not every exergonic reaction is necessarily exothermic i.e. the energy is not always released as heat, thus a reaction with ΔG<0 may have ΔH<0, ΔH=0 or ΔH>0. These characteristics were properly explained in Gibbs fundamental equation as follows:

$$\Delta G = \Delta H - T \Delta S \tag{1}$$

Gibbs equation represents a state function, which means that it solely depends on the equilibrium state of the system, indistinctly of how the system got to that state. However, catalytic mechanisms move in a tighter boundary of the chemical reaction process, they will not influence the initial nor the final state of the reaction, but will influence in the path leading from one to the other: The enzymatic process diminishes the required free energy for the reaction to take place (Figure 2). The occurrence of spontaneous chemical reactions is restricted by the influence of an energetic barrier, appointed as activation energy; in order for any reaction to take place, there must be enough energy provided to the system to outperform it into a transition state – in a non catalyzed reaction, this can be done e.g. by heating up the system –, this critical instant represents a state where the reactants and the products are simultaneously existent, due to the concurrence of bonds both from the reactant state and the product state, such molecular form is highly unstable, hence related to a large amount of free energy. The enzyme-catalyzed reaction will present an alternative path for the reaction to take place (Jencks, 1987; Copeland, 2004).

By understanding the catalytic process as an energy state transformation, it results easier to elucidate an expected outcome by analytical means without much regard on the details on a specific metabolic pathway. Lets consider again the Gibbs fundamental free energy equation under constant temperature conditions, as given in Equation 1. Consider as well an exergonic reaction, thus no energy is provided to the system, nor by external means neither as a result of a previous enzymatic reaction on the metabolic pathway: For the simplest case we can suppose that no energy was released from the system, the only possible outcome would be an increase on the entropy of the system. When the substrate is composed from molecules constructed by different chemical elements, a dissociation of the elements can be expected, in some cases at least one of the molecules can be readily detected through a known chemical-sensing method. If such is the case, it can be said that the sensing strategy would not be targeted to the substrate, but to a byproduct or a related reaction, this case will be referred as a Biotransformation-Based whole-cell sensing strategy (BtB Strategy) – A special situation will be considered when the result of such biotransformation yield a readily detection signal, such as bioluminescence, often addressed as bioreporter (Leveau & Lindow, 2002; Jansson, 2003; van der Meer & Belkin, 2010).

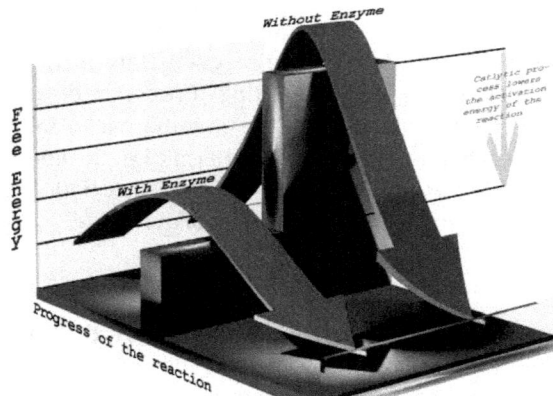

Figure 2. Free energy diagram through the progress of a reaction comparing a catalyzed reaction against a non-catalyzed reaction. The enzymatic catalyst would lower the activation energy, increasing the rate where a normally "spontaneous" reaction would take place

An additional case can be elucidated when no entropy change is considered, hence, the only effect on the resulting free energy equation would be induced by a change of enthalpy, if the initial conditions agree on those previously proposed, the reaction would be exothermic, the outcome is then a release of energy which can be measured by suitable methods. This approach will be labeled as a Stimuli-Response-Based whole-cell strategy (SRB Strategy).

The actual scenarios of the enzyme-catalyzed reaction are much more complicated than the above considered, the metabolic pathway sets a complex network of reactions that can be interrelated with various pathways, some of them active all-through the cell vital routines, other related to the specific social behavior of different strains, in cases for both intra- and inter-species responses (Waters & Bassler, 2005; Shank & Kolter, 2009). Such scenarios combine different characteristics on the entropic and enthalpic properties of the reaction, and the endergonic-exergonic nature of the reaction, nonetheless such combinations would lead to a similar analysis, favoring either the expression of byproducts, according to the BtB strategy, or by energetic measurable changes approached by the SRB strategy.

3. Whole-cell sensing methods, transduction

In the previous section it was introduced two biosensing strategies based on the specific contributory thermodynamic effects of each term, i.e. Stimuli-response-based strategy (SRB strategy), grounded on the concept of enthalpy (energy related response), and Biotransformation-based strategy (BtB strategy), linked to entropy related intake (molecular related response). Transduction is referred, in this section, as the process in which the presence of a given substrate triggers a suitable reaction to an information translated measurable unit.

The concept in this section is closer to that posed by the analytical procedure of the physical context, for both SRB and BtB strategies; the scope approached refers to the interfaced-measurable signal, anyway, few insights on the physics and thermodynamic background of some concepts are given, which are usually neglected in literature.

3.1. Stimuli-Response-Based strategy (SRB strategy)

The SRB strategy is assumed under the energy release/intake principle, the main scenario, presented at the introduction, would be the case of an exothermic reaction. Given this first glimpse, such strategy might be conceived of being based on the whole-cell heat production as a responsive event when interacting with a substrate, yet, even though the concept encloses direct reference to the thermal properties of the reaction, the release/intake of energy is not necessarily of thermal nature, effects on the electric domain are also to be considered. The reactions driven by the whole-cell enzymatic activity, for the consideration of an SRB strategy, rely upon the ionic effects, as are ionic transport phenomena (Borkholder, 1998; Ikeda & Kano, 2001), which provoke alterations over an adequate material, namely electrode, readily measurable on the electric domain.

The most general reaction in which biological systems are involved is known as reduction-oxidation (redox) reaction. As it deploys, reactants undergo an electron transfer process, which produces fractional quantities of Gibbs free energy (Marcus, 1956a; Marcus, 1956b), that are released in the form of heat; the breaking, formation and reconfiguration of atomic bonds contribute likewise to the production of heat, both effects reflect the metabolic activity of a given strain when interacting with specific substrates. Thermal changes respond directly to these chemical reactions (Newman & Thomas-Alyea, 2012), the energy released in such reactions is easily profited by the usage of temperature detection methods, applicable to biological interactions through microcalorimetric processes, early described by Max Rubner (Rubner, 1911), assessing the measure of heat flow of a biological process, which develops proportionally as chemical interactions take place (Braissant, Wirz, Göpfert & Daniels, 2010).

Akin interaction occurs on the boundaries of a whole-cell-electrode interface; the electrode serves as a suitable electron donor/acceptor for the redox reaction deployed on the presence of the substrate, the released ionic clusters move from a higher concentration region to a lower concentration region, a diffusion transport effect due to the concentration gradient (Newman & Thomas-Alyea, 2012); the cell potential was first explained by Nernst, who related such potential to the Gibbs free energy state of the reactants. (Nernst & Barr, 1926). For every chemical reaction, there is a free energy dependent change (Ulstrup & Jortner, 1975); the alterations induced on the electrode modify its electric structure and generate readily measurable electrical units, typically, such reaction achieves either a difference in current (amperometric), a potential or charge accumulation (potentiometric), alters the conductive properties between surfaces (conductometric), produces changes on the impedance (impedimetric) or potentiometric changes on a gate electrode (field-effect) (Ikeda & Kano, 2001; Thévenot et al., 2001; Grieshaber, MacKenzie, Vörös & Reimhult, 2008). An additional interaction for the considered reaction is posed by microbial fuel cells (MFC) biosensors (Aston & Turner, 1984; Stein, Keesman, Hamelers & van Straten, 2011), the resulting ions that are needed by the product of the whole-cell enzymatic reactions or that are detached from them can be exchanged through the whole-cell membrane, the ionic concentration gradient produces an electromotive force (EMF) measurable as an electric energy difference. However, by reason of the non-conductive nature of the membrane, not every cell posses the ability to directly transfer the ion cluster towards the electrode; a membrane is commonly used that would act as a mediator to selectively transfer electrons/protons to the electrode, yet, some electrochemically active strains (Chang, Moon & Bretschger, 2006), as *Aeromonas hydrophilia* (Pham, Jung, Phung, Lee, Chang, Kim et al., 2003), *Clostridium butyricom* (Park, Kim, Kim, Kim, Kim, Kim et al., 2001), *Desulfoblus propionicus* (Holmes, Bond & Lovley, 2004), *Enterococcus gallinarum* (Kim, Hyun, Chang, Kim, Park, Kim et al., 2005), *Geobacter sulfurreducens* (Bond & Lovley, 2003), *Rhodofoferax ferrireducens* (Chaudhuri & Lovley, 2003) and *Shewanella putrefaciens* (Kim, Park, Hyun, Chang, Kim & Kim, 2002), would be able to provide a mediator less MFC.

As the reaction is driven by the transformation of the substrate, the resultant electric gradient is proportional to the substrate's concentration; in consequence, the output signal is a quantitative indicator dependent on the amount of the specific substrate interacting with the whole-cell surface (Aston & Turner, 1984). Under the present construct, the detection proposed for the use of the SRB strategy is only mediated by the utilization of an electrode, directly interfaced with the whole-cell or mediated through an electron/proton exchange membrane.

For any of the exposed cases, the SRB strategy is considered when the interaction with a given substrate generates a measurable change over an electrode interfacing the whole-cell, these methods usually require the immobilization of the whole-cell on the surface of the electrodes, techniques for cell entrapment as polyvinyl alcohol (PVA) immobilization (Rouillon,, Tocabens & Carpentier, 1999), hydrogel immobilization (Gäberlein, Spener & Zaborosch, 2000), immobilization crosslinking method (Babu, Patra, Karanth, Kumar & Thakur, 2007), physical confinement (Hernandez, Gaviria, Segura & Osma, 2013), among others.

There are two characteristics shared by the methods arranged under this strategy, the prevalence of the contribution of the enthalpy of the system on the overall reaction and the direct usage of electrodes to assess a proportional electrical measure. The consideration to use an SRB strategy with a suitable whole-cell strain is advantageous when a one reaction one response measures is required as a non mediated direct response, which is translated in faster acquisition of the measure, less complex systems for data transduction and lesser probability of side effect-reactions to add noise to the measure.

3.2. Biotransformation based strategy (BtB strategy)

BtB strategy is considered, under the present scope, as a set of byproduct mediated-sensing methods. Many reaction results can be contemplated strictly within the chemical realm of the dissociation and consumption of different compounds. The transformation occurred within the whole-cell can be readily detected by different methods established for precise substances and not by the energy produced directly within the whole-cell. Some examples of the used electrodes are listed in Table 4.

The conformation of chemical bonds, from the perspective of the molecular interaction, derived from specific metabolic pathways, serves as a further option for targeting a substrate. The processes delivered by the whole-cell might be only possible if complemented with a different substance, e.g. aerobic organisms depend on the consumption of oxygen to complete the processes within their complete metabolic network, the amount of oxygen on a contained environment can be used as a measure unit of the whole-cell activity. Further applications related to the produced substances due to whole-cell respiration products and effects, such as the detection on pressure change and CO_2 (Lei, Chen & Mulchandani, 2006), other products due to the separation of compounds that can be easily assessed can be profited.

Transducer	Species detected
Amperometric electrodes	O_2, H_2O, NADH, I_2
Ion-selective electrodes	H^+, NH_4^+, NH_3, CO_2, I^-, CN^-
Field-effect transistors	H^+, H_2, NH_3
Photomultiplier (in conjunction with fiber optics)	Light emission or chemiluminescence
Photodiode (in conjunction with a light-emitting diode)	Light absorption
Piezoelectric crystal	Mass adsorbed

Table 4. Some transducers used under the scope of the BtB strategy (indirect electrodes), table extracted from (Aston & Turner, 1984)

The transduction for a BtB strategy would require further steps beyond the strain-specific reactions; the biological task is limited to the production of intermediate agents to unleash further chemical reactions that could be translated e.g. into electrical signals, chemical analogous to the methods presented for the transduction on SRB strategy. The recognition and quantification of the produced byproduct must comply with simple procedures.

The targeting of the production and intake of substances for reporting the presence of a given substrate might be addressed as the utilization of bioreporters (Daunert et al., 2000), identified by means of the specific reporter gene responsible to manage the production of the substance on a specific metabolic pathway. The term bioreporter, is usually referred in regards to very specific products that can be detected e.g. by means of optical instruments based on fluorescence and phosphorescence (Pringsheim, 1949).

Bioreporter genes can be introduced within the cell genome by different methods leveraged from genetic engineering (Salis et al., 2009; van der Meer & Belkin, 2010), they are attached to precise metabolic pathways, due to the linear nature of the deployment of reactions within the metabolic pathways, the bioreporter would only be active when the correspondent pathway has reached the stage in which the specific reporter gene is reached, thus reporting on the presence of the sought substrate.

The released molecules become the target of chemical sensing methods, bridging the translation for the quantification of the substrate, in the same way the electrical response on the SRB method is proportional to the amount of substrate detected, this molecules would be released/consumed as there is more/less concentration of the substrate.

4. Current advances in whole-cell biosensors

The selection of a suitable strategy for the development of a biosensor responds to different properties rendered by the selection of the used strain. One of the main challenges is to recognize the specific substrate which can be targeted by different strains, or which strain best adapts to a specific target. In Table 5, an extensive review on 163 cases is made, compiling 116 different whole-cell strains, differentiating the reported genetically modified ones, and a record of about 100 different targets that have been claimed to be recognized for such strains.

Microorganism	Target	Detection method	Strategy	Reference
A. aceti (IFO 3284)	Ethanol	Amperometric	SRB	(Ikeda, Kato, Maeda, Tatsumi, Kano & Matsushita, 1997)
A. adeninivorans LS3	BOD	Amperometric	SRB	(Riedel, Lehmann, Tag, Renneberg & Kunze, 1998; Chan, Lehmann, Tag, Lung, Riedel, Gruendig et al.,1999; Chan, Lehmann, Chan, Chan, Chan, Gruendig et al., 2000; Tag, Kwong, Lehmann, Chan, Renneberg, Riedel et al., 2000; Tag, Lehmann, Chan, Renneberg, Riedel & Kunze, 2000)
A. ferrooxidans	Fe^{2+} $S_2O_3^{2-}$ $Cr_2O_7^{2-}$	mperometric	SRB	(Zlatev, Magnin, Ozil, & Stoytcheva, 2006a; Zlatev, Magnin, Ozil, & Stoytcheva, 2006b; Zlatev, Magnin, Ozil, & Stoytcheva, 2006c)
A. globiformis	Choline	Amperometric	SRB	(Stoytcheva, Zlatev, Valdez, Magnin & Velkova, 2006)
Arthrobacter. sp. JS 443	p-Nitrophenol	Amperometric	SRB	(Lei, Mulchandani, Chen, Wang & Mulchandani,2003; Lei, Mulchandani, Chen, Wang & Mulchandani, 2004)
A. nicotianae (acyl-CoA oxidase)	Short chain fatty acids in milk (butyric acid)	Oxygen electrode (Polyvinyl alcohol)	BtB	(Ukeda, Wagner, Bilitewski & Schmid, 1992; Ukeda, Wagner, Weis, Miller, Klostermeyer & Schmid, 1992; Schmidt, Standfuß-Gabisch & Bilitewski, 1996)
A. niger	Ethanol	Amperometric	SRB	(Subrahmanyam, Shanmugam, Subramanian, Murugesan, Madhav & Jeyakumar, 2001)
	Glucose	Oxygen electrode (entrapment Glucose) in dialysis membrane)	BtB	(Katrlík, Švorc, Rosenberg & Miertuš, 1996)
A. peroxydans	Hydrogen peroxide	Amperometric	SRB	(Sumathi, Rajasekar & Narasimham, 2000)
A. phenologenes (Tyrosine-phenol lyase)	Tyrosine	NH3 gas sensing electrode (direct immobilisation on sensor membrane)	BtB	(Di Paolantonia & Rechnitz, 1982)
Activated sludge	BOD	Amperometric	SRB	(Kumlanghan, Kanatharana, Asawatreratanakul, Mattiasson & Thavarungkul, 2008)
Activated sludge (mixed microbial consortium)	BOD	Oxygen electrode/flow injection system (entrapped in dialysis membrane)	BtB	(Liu, Björnsson & Mattiasson, 2000)
Anaerobic sludge	BOD	MFC	SRB	(Di Lorenzo, Curtis, Head & Scott, 2009)
B. ammoniagenes	Urea	Conductometric	SRB	(Jha, Kanungo, Nath & D'Souza, 2009)

Microorganism	Target	Detection method	Strategy	Reference
B. stearothermophilus var. calidolactis	β-lactams	Potentiometric	SRB	(Ferrini, Mannoni, Carpico & Pellegrini, 2008)
B. subtilis	Peptides (aspartame)	Oxygen electrode (filter paper strip and dialysis membrane)	BtB	(Renneberg, Riedel & Scheller, 1985)
	Enalapril maleate (angiotensin)	Oxygen electrode		(Fleschin, Bala, Bunaciu, Panait & Aboul-Enein, 1998)
	BOD	Amperometric	SRB	(Riedel, Renneberg, Kühn & Scheller, 1988)
B. subtilis (heat killed)	BOD	Amperometric	SRB	(Tan & Qian, 1997)
B. subtilis and *B. licheniformis 7B*	BOD	Amperometric	SRB	(Tan, Li, Neoh & Lee, 1992; Tan, Li & Neoh, 1993; Li, Tan & Lee, 1994)
Bacillus sp.	Urea	NH_4^+ ion selective electrode	BtB	(Verma & Singh, 2003)
Bacteria consortium	Glucose	MFC	SRB	(Kumlanghan, Liu, Thavarungkul, Kanatharana & Mattiasson, 2007)
Bioluminescent recombinant *E. colir::luxAB* strain	Tributyltin	Luminescence	BtB	(Thouand, Horry, Durand, Picart, Bendriaa, Daniel et al., 2003)
BOD-Multiple organisms	BOD	Amperometric	SRB	(Tan & Wu, 1999)
Brevibacterium sp.	Acrylamide; acrylic acid	Oxygen electrode (free cells)	BtB	(Ignatov, Rogatcheva, Kozulin & Khorkina, 1997)
C. parapsilosis	BOD	Amperometric	SRB	(König, Reul, Harmeling, Spener, Knoll & Zaborosch, 2000)
C. sp.	Cu^{2+}	Voltammetric	SRB	(Alpat, Cadirci, Yasa & Telefoncu, 2008)
C. tropicalis	Ethanol	Amperometric	SRB	(Akyilmaz & Dinçkaya, 2005)
Candida vini	Alcohol	Oxygen electrode (porous acetyl cellulose filter)	BtB	(Mascini, Memoli & Olana, 1989)
Chorella vulgaris (algae)	Atrazine	Amperometric	SRB	(Shitanda, Takamatsu, Watanabe & Itagaki, 2009)
	Cd^{2+}			(Guedri & Durrieu, 2008)
	Cd^{2+} and Zn^{2+} Paraoxon-methyl	Conductometric		(Chouteau, Dzyadevych, Durrieu & Chovelon, 2005)
	Phosphate	Oxygen electrode (polycarbonate membrane)	BtB	(Matsunaga, Suzuki & Tomoda, 1984)
Chloroplast /thylakoid membranes	Herbicides[c] (diuron and atrazine)	Pt-electrode in microelectrochemical cell (photo cross linkable PVA bearing styrylpyridium group	BtB	(Rouillon, Sole, Carpentier & Marty, 1995)

Microorganism	Target	Detection method	Strategy	Reference
CO_2 utilizing autotropic bacteria (*Pseudomonas*)	CO_2	Oxygen electrode (bound on cellulose nitrate membrane)	BtB	(Suzuki, Tamiya & Karube, 1987)
Comamonas testosterone TI	Non-ionic surfactants	Amperometric	SRB	(Taranova, Fesay, Ivashchenko, Reshetilov, Winther-Nielsen & Emneus, 2004)
E. coli	Vitamin B-12	Oxygen electrode (trapped in porous acetyl cellulose membrane)	BtB	(Karube, Wang Tamiya, & Kawarai, 1987)
	Phenol	Amperometric	SBR	(Neufeld, Biran, Popovtzer, Erez, Ron & Rishpon, 2006)
	Nalidixic acid			(Ben-Yoav, Biran, Pedahzur, Belkin, Buchinger, Reifferscheid et al., 2009)
E. coli bearing *fab A'::lux* fusions	Pollutants/toxicity	Luminescence	BtB	(Bechor, Smulski, Van Dyk, LaRossa & Belkin, 2002)
E. coli DH5α (pPR-arsR-ABS, expressing *egfp*)	Arsenite	Fluorescence	BtB	(Wells, Gösch, Rigler, Harms, Lasser & van der Meer, 2005)
E. coli DPD1718 containing *recA'::lux fusion*	Genotoxicants	Luminescence	BtB	(Polyak, Bassis, Novodvorets, Belkin & Marks, 2000)
E. coli HB101 pUCD607 containing *luxCDABE* cassette	Water pollutants/toxicity	Luminescence	BtB	(Horsburgh, Mardlin, Turner, Henkler, Strachan, Glover et al., 2002)
E. coli HMS174 harboring *mer-lux* plasmid *pRB27*	Bioavailable mercury	Luminescence	BtB	(Rasmussen, Turner & Barkay, 1997)
E. coli HMS174 harboring *mer-lux* plasmid *pRB27 or pRB28*	Hg^{2+}	Luminescence	BtB	(Rasmussen, Sørensen, Turner & Barkay, 2000)
E. coli HMS174 harboring *mer-lux* plasmid *pRB27 or pRB28*	Bioavailable mercury	Luminescence	BtB	(Barkay, Gillman & Turner, 1997)
E. coli HMS174 harboring *mer-lux* plasmid *pRB28*, *pOS14orpOS15*	Bioavailable mercury	Luminescence	BtB	(Selifonova, Burlage & Barkay, 1993)
E. coli K12	Mono-and /disaccharides	Amperometric	SRB	(Held, Schuhmann, Jahreis & Schmidt, 2002)
E. coli MC1061 harboring *mer-lux* plasmid *pTOO11*	Urinary mercury (II)	Luminescence	BtB	(Roda, Pasini, Mirasoli, Guardigli, Russo, Musiani et al., 2001)
E. coli MC4100 harboring *pAHL-GFP*	N-Acyl homoserine lactones in soil	Fluorescence	BtB	(Burmølle, Hansen, Oregaard & Sørensen, 2003)
E. coli WP2	Tryptophan	LAPS	SBR	(Seki Kawakubo, Iga & Nomura, 2003)

Microorganism	Target	Detection method	Strategy	Reference
E. coli, Pantoea agglomerans and *Pseudomonas syringae*	Water availability	Fluorescence	BtB	(Axtell & Beattie, 2002)
F. solani	Acetic acid	Amperometric	SRB	(Subrahmanyam, Kodandapani, Shanmugam, Moovarkumuthalvan, Jeyakumar & Subramanian, 2001)
Flavobacteium sp.	Organophosphates	pH electrode	SRB	(Gäberlein et al., 2000)
G. oxydans	Ethanol	Amperometric	SRB	(Tkáč, Voštiar, Gemeiner & Šturďik, 2002; Tkac, Vostiar, Gorton, Gemeiner & Sturdik, 2003)
	Total sugars			(Tkáč, Gemeiner, Švitel, Benikovský, Šturďik, Vala et al., 2000)
	Glucose			(Odaci, Timur & Telefoncu, 2009)
				(Tuncagil, Odaci, Varis, Timur & Toppare, 2009; Tuncagil, Odaci, Yildiz, Timur & Toppare, 2009)
	Ethanol			(Valach, Katrlík, Šturdík & Gemeiner, 2009)
	1,3-Propanediol			(Katrlík, Vostiar, Sefcovicová, Tkác, Mastihuba, Valach et al., 2007)
G. oxydans (D-glucose dehydrogenase), *S.cerevisiae* (invertase), *K. marxianus* (galactosidase)	Glucose, sucrose, lactose	Oxygen electrode (gelatine)	BtB	(Svitel, Curilla & Tkác, 1998)
G. oxydans or *P. methanolica*	Ethanol	Amperometric	SRB	(Reshetilov, Trotsenko, Morozova, Iliasov & Ashin, 2001)
G. suboxydans	Ethanol	Amperometric	SRB	(Kitagawa, Ameyama, Nakashima, Tamiya & Karube, 1987)
G. sulfurreducens	Acetate	MFC	SRB	(Tront, Fortner, Plötze, Hughes & Puzrin, 2008)
GEfMb E. coli (organophosphorous hydrolase)	Organophosphate nerve agents (paraxon, methyl parathion, diazinon)	Potentiometric (adsorption on electrode surface)	SRB	(Mulchandani, Mulchandani, Kaneva & Chen, 1998)
		Fiber-optic (agarose)	BtB	(Mulchandani, Kaneva & Chen, 1998)
H. polymorpha	L-lactate	Amperometric	SRB	(Smutok, Dmytruk, Gonchar, Sibirny & Schuhmann, 2007)
K. oxytoca AS1	BOD	Amperometric	SRB	(Ohki, Shinohara, Ito, Naka, Maeda, Sato et al., 1994)
LAS degrading bacteria isolated from activated sludge	Aniomic surfactants (linear alky benzene sulfonates- LAS)	Oxygen electrode, (reactor type sensor, ca-alginate)	BtB	(Nomura, Ikebukuro, Yokoyama, Takeuchi, Arikawa, Ohno et al., 1994)

Microorganism	Target	Detection method	Strategy	Reference
Moraxella sp.	β-d-Glucuronidase	Amperometric	SRB	(Togo, Wutor, Limson & Pletschke, 2007)
	p-Nitrophenol			(Mulchandani, Lei, Chen, Wang & Mulchandani, 2002; Mulchandani, Hangarter, Lei, Chen & Mulchandani, 2005)
	Paraoxon			(Mulchandani, Chen & Mulchandani, 2006)
Microbial consortium	BOD	Amperometric	SRB	(Rastogi, Kumar, Mehra, Makhijani, Manoharan, Gangal et al., 2003; Liu, Olsson & Mattiasson, 2004a; Liu, Olsson & Mattiasson, 2004b; Dhall, Kumar, Joshi, Saxsena, Manoharan, Makhijani et al., 2008)
Nitrobacter vulgaris DSM10236	Nitrite	Oxygen electrode (adsorption on Whatman paper)	BtB	(Reshetilov, Iliasov, Knackmuss & Boronin, 2000)
P. aeruginosa	Cephalosporins	Potentiometric	SRB	(Kumar, Kundu, Pakshirajan & Dasu, 2008)
P. aeruginosa + K. sp.	Methane	Amperometric	SRB	(Wen, Zheng, Zhao, Shuang, Dong & Choi, 2008)
P. aeruginosa FRD1 carrying plasmid pMOE15 with recA::luxCDABE	UV	Luminescence	BtB	(Elasri & Miller, 1999)
P. aeruginosa JI104	Trichloroethylene	Chloride ion selective electrode	BtB	(Han, Kim, Sasaki, Yano, Ikebukuro, Kitayama et al., 2001; Han, Sasaki, Yano, Ikebukuro, Kitayama, Nagamune et al., 2002)
P. alcaligenes	Caffeine	Amperometric	SRB	(Babu et al., 2007)
P. angusta	Ethanol	Amperometric	SRB	(Voronova, Iliasov & Reshetilov, 2008)
P. fluorescens	BOD	Amperometric	SRB	(Yoshida, Yano, Morita, McNiven, Nakamura & Karube, 2000; Yoshida, Hoashi, Morita, McNiven, Nakamura & Karube, 2001)
	Glucose			
	Galactose			(Kirgoz, Timur, Odaci, Pérez, Alegret & Merkoçi, 2007; Odaci, Kiralp Kayahan, Timur & Toppare, 2008; Yeni, Odaci & Timur, 2008; Tuncagil, Odaci, Varis, Timur & Toppare, 2009)
	Mannose			
	Xylose			(Odaci, Timur & Telefoncu, 2008)
P. fluorescens 10586r pUCD607	Toxicity of chlorophenol	Luminescence	BtB	(Tiensing, Strachan & Paton, 2002)
P. fluorescens A506 (pTolLHB) and E. cloacae JL1157 (pTolLHB)	Bioavailable toluene and related compounds	Fluorescence	BtB	(Casavant, Thompson, Beattie, Phillips & Halverson, 2003)

Microorganism	Target	Detection method	Strategy	Reference
P. fluorescens DF57 with a *Tn5::luxAB* promoter probe transposon	Bioavailable copper	Luminescence	BtB	(Tom-Petersen, Hosbond & Nybroe, 2001)
P. fluorescens pUCD607	Pollution-induced stress	Luminescence	BtB	(Porteous, Killham & Meharg, 2000)
P. fluorescents NCIMB 11764	Cyanide	Amperometric	SRB	(Lee & Karube, 1996)
P. putida	BOD	Oxygen electrode (adsorption on porous nitro cellulose membrane)	BtB	(Chee, Nomura & Karube, 1999)
	Phenolic compounds	Oxygen electrode (reactor with cells adsorbed on PEI glass)		(Nandakumar & Mattiasson, 1999b)
	3-Chloro-benzoate	Oxygen electrode (PVA)		(Riedel, Naumov, Boronin, Golovleva, Stein & Scheller, 1991)
	Phenolic compounds	Amperometric	SRB	(Timur, Pazarlioğlu, Pilloton & Telefoncu, 2003; Timur, Della Seta, Pazarlioğlu, Pilloton & Telefoncu, 2004)
	Phenol			(Kirgöz, Odaci, Timur, Merkoçi, Pazarlioğlu, Telefoncu et al., 2006)
	Galactose			(Timur, Anik. Odaci & Gorton, 2007)
	Glucose			
	Catechol			(Timur, Haghighi, Tkac, Pazarlioğlu, Telefoncu & Gorton, 2007)
	2,4-Dichloro phenoxy acetic acid			(Odaci, Sezgintürk, Timur, Pazarlioğlu, Pilloton, Dinçkaya et al., 2009)
P. putida carrying *NAH7* plasmid and a chromosomally inserted gene fusion between the *sal* promoter and the *luxAB* genes	Bioavailable naphthalene	Luminescence	BtB	(Werlen, Jaspers & van der Meer, 2004)

Microorganism	Target	Detection method	Strategy	Reference
P. putida F1	Benzene	Amperometric	SRB	(Rasinger, Marrazza, Briganti, Scozzafava, Mascini & Turner, 2005)
	Toluene			
	Ethylbenzene			
P. putida JS444	Paraoxon	Amperometric	SRB	(Lei, Mulchandani, Chen & Mulchandani, 2005; Lei, Mulchandani, Chen, Wang & Mulchandani, 2005)
	Parathion			
	Methyl parathion			
	Fenitrothion			(Lei, Mulchandani, Chen & Mulchandani, 2006; Lei, Mulchandani, Chen & Mulchandani, 2007)
	EPN			
P. putida SG10	BOD	Amperometric	SRB	(Chee, Nomura, Ikebukuro & Karube, 2005)
Pseudomonas. sp.	Microbiologically influenced corrosion	Amperometric	SRB	(Dubey & Upadhyay, 2001; Banik, Prakash & Upadhyay, 2008)
	p-Nitrophenol			
	BOD			(Li & Chu, 1991)
P. syringae	BOD	Amperometric	SRB	(Kara, Keskinler & Erhan, 2009)
P. vulgaris (Phenylalanine deaminase)	Phenylalanine	Amperometric oxygen electrode	BtB	(Liu, Cui & Deng 1996)
P.fischeri and *P. putida BS566::luxCDABE*	Toxicity of waste water treatment plant treating phenolics-containing waster	Luminescence	BtB	(Philp, Balmand, Hajto, Bailey, Wiles, Whiteley et al., 2003)
Potato (*S. tuberosum*) slices (polyphenol oxidase inhibition)	Mono and polyphenols (atrazine)	Oxygen electrode (tissue slice sandwitched between membranes)	BtB	(Mazzei, Botrè, Lorenti, Simonetti, Porcelli, Scibona et al., 1995)
Pseudomonas and *Archromobacter*	Anionic surfactants	Amperometric	SRB	(Taranova, Semenchuk, Manolov, Iliasov & Reshetilov, 2002)
Psychrophilic *D. radiodurans*	Sugars (glucose)	Oxygen electrode (agarose)	BtB	(Nandakumar & Mattiasson, 1999a)
R. erthropolis	2,4-Dinitrophenol	Amperometric	SRB	(Emelyanova & Reshetilov, 2002)
Ralstonia eutropha AE2515	Ni^{2+} and Co^{2+}	Luminescence	BtB	(Tibazarwa, Corbisier, Mench, Bossus, Solda, Mergeay et al., 2001)
Recombinant *E. coli*	Penicillin	Flat pH electrode	BRB	(Galindo, Bautista, García & Quintero, 1990; Chao & Lee, 2000)
	Cadmium	Amperometric		(Biran, Babai, Levcov, Rishpon & Ron, 2000)
Recombinant *E. coli* containing DL-2-haloacid dehalogenase encoding gene and *luxCDABE* genes	Halogenated organic acids	Luminescence	BtB	(Tauber, Rosen & Belkin, 2001)

Microorganism	Target	Detection method	Strategy	Reference
Recombinant *E. coli* containing *recA'::lux* fusion	UV	Luminescence	BtB	(Rosen, Davidov, LaRossa & Belkin, 2000)
Recombinant *Moraxella*	Organophosphates	Amperometric	SRB	(Mulchandani, Chen, Mulchandani, Wang & Chen, 2001)
Recombinant *P. putida JS 444*	Organophosphates	Amperometric	SRB	(Lei, Mulchandani, Chen & Mulchandani, 2005)
Recombinant *Pseudomonas syringae* carrying *gfp* gene	Bioavailable iron	Fluorescence	BtB	(Joyner & Lindow, 2000)
Recombinant *S. cerevisiae*	Cu^{2+}	Amperometric	SRB	(Lehmann, Riedel, Adler & Kunze, 2000)
Rhodococcus erythropolis DSM Nr. 772 and *Issatchenkia orientalis DSM Nr. 3433*	BOD	Amperometric	SRB	(Heim, Schnieder, Binz, Vogel & Bilitewski 1999)
Rhodococcus sp. DSM 6344	Chlorinated and brominated hydrocarbons (1-chlorobutane and ethylenebromide)	Ion selective electrodes (alginate)	BtB	(Peter, Hutter, Stöllnberger& Hampel, 1996)
Rhodococcus sp.; Trichosporon beigelii	Chlorophenols	Oxygen electrode (PVA)	BtB	(Riedel, Hensel, Rothe, Neumann & Scheller, 1993; Riedel, Beyersdorf-Radeck, Neumann & Schaller, 1995)
S. cerevisiae	Sucrose	Amperometric	SRB	(Rotariu, Bala & Magearu, 2000)
	Cyanide	Oxygen electrode (PVA)	BtB	(Ikebukuro. Honda, Nakanishi, Nomura, Masuda, Yokoyama et al., 1996; Nakanishi, Ikebukuro & Karube, 1996; Ikebukuro, Miyata, Cho, Nomura, Chang, Yamauchi et al., 1996)
	BOD	Amperometric	SRB	(Nakamura, Suzuki, Ishikuro, Kinoshita, Koizumi, Okuma et al., 2007)
	Vitamin B1	Amperometric	SRB	(Akyilmaz, Yaşa & Dinçkaya, 2006)
	L-lysine	Amperometric	SRB	(Akyilmaz. Erdoğan, Oztürk & Yaşa, 2007)
	Sucrose	Oxygen	BtB	(Rotariu, Bala & Magearu, 2002)
S. cerevisiae (I) (II)	Cu^{2+}	Amperometric	SRB	(Tag, Riedel, Bauer, Hanke, Baronian & Kunze, 2007)
S. ellipsoideus	Ethanol	Amperometric	SRB	(Rotariu & Bala, 2003)
S. ellipsoideus	Ethanol	Oxygen	BtB	(Rotariu, Bala & Magearu, 2004)
S. typhimurium	2-Amino-3-methylimidazo[4,5-f]quinoline	Amperometric	SRB	(Ben-Yoav, Elad, Shlomovits, Belkin & Shacham-Diamand, 2009)

Microorganism	Target	Detection method	Strategy	Reference
S. uvarum	Vitamin B-6	Oxygen electrode (adsorption on cellulose nitrate membrane)	BtB	(Endo, Kamata, Hoshi, Hayashi & Watanabe, 1995)
Salt tolerant mycelial yeast *A. adeninivorans* LS3	BOD	Oxygen electrode (PVA)	BtB	(Tag, Lehmann, Chan, Renneberg, Riedel & Kunze, 1998)
Serratia marcescens LSY4	BOD	Amperometric	SRB	(Kim & Kwon, 1999)
Sinorhizobium meliloh containing a *gfp* gene fused to the *melA* promoter	Galactosides	Fluorescence	BtB	(Bringhurst, Cardon & Gage, 2001)
Sphingomonas yanoikuyae B1 or *Ps. fluorescens* WW4	Polycyclic aromatic hydrocarbons (Naphthalene)	Oxygen electrode (polyurethane based hydrogel)	BtB	(König, Zaborosch, Muscat, Vorlop & Spener, 1996; König, Zaborosch & Spener, 1997)
Streptococcus faecium (Pyruvate dehydrogenase complex)	Pyruvate	CO2 gas sensing electrode (direct immobilisation on sensor membrane)	BtB	(Di Paolantonia & Rechnitz, 1983)
Synechococcus PCC 7942 reporter strain	Bioavailable phosphorus	Luminescence	BtB	(Schreiter, Gillor, Post, Belkin, Schmid & Bachmann, 2001)
Synechococcus sp. PCC 7942	Pollutants such as diuron and mercuric chloride	Photoelectrochemical (photo cross linkable PVA bearing styrylpyridium group)	BtB	(Rouillon et al., 1999)
T. bacteria	BOD	Amperometric	SRB	(Karube, Yokoyama, Sode & Tamiya, 1989)
T. candida	BOD	Amperometric	SRB	(Sangeetha, Sugandhi, Murugesan, Murali Madhav, Berchmans, Rajasekar et al., 1996)
T. cutaneum	BOD	Oxygen electrode array (Miniature electrode photo cross-linkable resin) (Entrapment)	BtB	(Marty, Olive & Asano, 1997; Yang, Sasaki, Karube & Suzuki, 1997)
T. cutaneum and *B. subtilis*	BOD	Amperometric	SRB	(Jia, Tang, Chen, Qi & Dong, 2003)
T. ferrooxidans	Cyanide	Amperometric	SRB	(Okochi, Mima, Miyata, Shinozaki, Haraguchi, Fujisawa et al., 2004)

Microorganism	Target	Detection method	Strategy	Reference
Trichosporum cutaneum	BOD	Miniature oxygen electrode (UV cross-linking resin ENT-3400)	BtB	(Yang, Suzuki, Sasaki & Karube, 1996)
Yeast	BOD	Amperometric	SRB	(Chen, Cao, Liu, & Kong, 2002)
Yeast cells	Bioavailable organic carbon n oxic sediments	Oxygen electrode (PVA)	BtB	(Neudörfer & Meyer-Reil, 1997)
Yeast SPT1 and SPT2	BOD	Amperometric	SRB	(Trosok, Driscoll & Luong, 2001)

Table 5. Different whole-cell strains for use in biosensing, it is listed the target, detection method and attributed strategy. Table modified from (D'Souza 2001; Lei, Chen et al., 2006; Su et al., 2011)

From the reviewed literature, approximately 51% of the strains respond to an SRB strategy, 45% to a BtB strategy and a 4% to both, 19 different methods are listed, 7 of them attributed to a SRB strategy, namely amperometric, conductometric, pH electrode, LAPS, MFC, potentiometric and voltametric, and the remaining 12 to the BtB strategy, comprising different chemical-compound-selective electrodes, and fluorescent and luminescent bioreporters.

The given classification is based on the nature of the different electrodes, if any, and aims to serve as a base for the correct identification of a specific strategy according to the application goals of the designing process of a biosensor. According to the U.S. Environmental Protection Agency (EPA), whole-cell biosensors display a promising alternative to the usage in early warning screening due to their fast reaction to toxins (EPA, 2005); in the same report, some technologies are disclosed under the current commercial application, although more accordingly with the definition of biological test, such systems are ToxScreen-II(currently III) (EPA, 2006; CheckLight Highlighting Water Safety-TOX-SCREEN 3, 2013), BioTox™ (EPA, 2006), DeltaTox® (EPA, 2003a), ToxTrak™ (Environmental technology verification program, 2006; ToxTrakTM Toxicity Reagent Set, 25-49 tests- Overview | Hach, 2013), POLYTOX™ (EPA, 2003b; InterLab Supply-Products-Biological Oxygen Demand (BOD) and Toxicity Testing Technology, 2013) and microMAX-TOX, the latter a promising device announced by the italian company Systea S.p.a. (SYSTEA S.p.A., 2013) which would cover the expected properties of a whole-cell based biosensor for online and continuous monitoring. Other commercially available whole-cell based biosensors are related in Table 6.

It is clearly observed the current preference to the utilization of the BtB strategy for commercial use, although the efforts posed by researchers to develop biosensors on the margins of the SRB strategy. The trending posed by the current developments under the scope of the SRB strategy will lead a new generation of biosensors based on the possibilities of different whole-cells strains to modify the electric structures of a given electrode, furthermore, the advantages regarding the needless utilization of reagents and the faster response will definitively play an important role on the favoring of inclusion of the development of SRB strategy biosensors.

The future developments would include the utilization of Archaea as an auspicious prospect on the development of highly effective SRB strategy biosensors, the affinity of such domain-type whole-cells with different substrates and the possibility it offers for strong electrode reactions is an interesting field to be explored.

Technology name	Biological system	Strategy	Reference
Aquasentinel	Algae	BtB	(Aqua Sentinel, 2013)
Fluotox	Algae (*Scendedesmus subspicatus*)	BtB	(Fluotox, 2013)
Lumitox	Algae (*Pyrocystis lunula*)	BtB	(Stiffey & Nicolaids, 1995)
Amtox	Bacteria	BtB	(Upton & Pickin, 1996)
Baroxymeter	Bacteria	BtB	(Baroxymeter, 2013)
BioTox Flash Test	Bacteria (*Vibrio fischerei*)	BtB	(Aboatox Environmental analysis, 2013)
Cellsense	Bacteria, *algae*	SRB	(Farré et al., 2001)
GreenScreen EM	Yeast (*Saccharomyces cerevisiae*)	BtB	(Keenan, Knight, Billinton, Cahill, Dalrymple, Hawkyard et al., 2007)
LUMIStox	Bacteria (*Vibrio fischerei*)	BtB	(Hach-Lange UK-LUMIStox, 2013)
MetPlate	Bacteria (*Escherichia coli*)	BtB	(MetPLATETM, 2013)
Sinorhizobium mellioti Toxicity Test	Bacteria (*Sinorhizobium mellioti*)	BtB	(van der Schalie, James & Gargan, 2006)

Table 6. Current commercially available whole-cell based biosensor, table modified from extract (Walther & Wurster, 2007)

References

Aboatox Environmental analysis (2013). Accessed August 31st. http://www.aboatox.com/environmental_analysis.html#flash.

Acha, V., Andrews, T., Huang, Q., Sardar, D.K., & Hornsby, P.J. (2010). Tissue-based bionsensors. In M. Zourob (Ed.). *Recognition Receptors in Biosensors*. New York, NY: Springer p.365-381. http://dx.doi.org/10.1007/978-1-4419-0919-0

Akyilmaz, E., & Dinçkaya, E. (2005). An amperometric microbial biosensor development based on Candida tropicalis yeast cells for sensitive determination of ethanol. *Biosensors & Bioelectronics, 20,* 1263-1269. http://dx.doi.org/10.1016/j.bios.2004.04.010

Akyilmaz, E., Erdoğan, A., Oztürk, R., & Yaşa, I. (2007). Sensitive determination of L-lysine with a new amperometric microbial biosensor based on Saccharomyces cerevisiae yeast cells. *Biosensors & Bioelectronics, 22,* 1055-1060. http://dx.doi.org/10.1016/j.bios.2006.04.023

Akyilmaz, E., Yaşa, I., & Dinçkaya, E. (2006). Whole cell immobilized amperometric biosensor based on Saccharomyces cerevisiae for selective determination of vitamin B1 (thiamine). *Analytical Biochemistry, 354,* 78-84. http://dx.doi.org/10.1016/j.ab.2006.04.019

Alpat, S., Cadirci, B., Yasa, I., & Telefoncu, A. (2008). A novel microbial biosensor based on Circinella sp. modified carbon paste electrode and its voltammetric application. *Sensors and Actuators B: Chemical, 134,* 175-181. http://dx.doi.org/10.1016/j.snb.2008.04.044

Aqua Sentinel (2013). Accessed August 31st. http://www.envirotechinstruments.com/aquasentinel.html.

Arnold, M. (1986). Potentiometric sensors using whole tissue sections. *Ion-Selective Electrode Rev., 8,* 85-113.

Arnold, M.A., & Rechnitz, G.A. (1980). Comparison of bacterial, mitochondrial, tissue, and enzyme biocatalysts for glutamine selective membrane electrodes. *Analytical Chemistry, 52,* 1170-1174. http://dx.doi.org/10.1021/ac50058a004

Asphahani, F., & Zhang, M. (2007). Cellular impedance biosensors for drug screening and toxin detection. *The Analyst, 132,* 835-841. http://dx.doi.org/10.1039/b704513a

Astor, W.J., & Turner, A.P.F. (1984). Biosensors and Biofuel Cells. *Biotechnology and Genetic Engineering Reviews, 1,* 89-120. http://dx.doi.org/10.1080/02648725.1984.10647782

Axtel, C.A., & Beattie, G.A. (2002). Construction and characterization of a proU-gfp transcriptional fusion that measures water availability in a microbial habitat. *Applied and Environmental Microbiology, 68,* 4604-4612. http://dx.doi.org/10.1128/AEM.68.9.4604-4612.2002

Babu, V.R.S., Patra, S., Karanth, N.G., Kumar, M.A., & Thakur, M.S. (2007). Development of a biosensor for caffeine. *Analytica chimica acta, 582,* 329-334. http://dx.doi.org/10.1016/j.aca.2006.09.017

Banerjee, P., & Bhunia, A.K. (2009). Mammalian cell-based biosensors for pathogens and toxins. *Trends in Biotechnology, 27,* 179-188. http://dx.doi.org/10.1016/j.tibtech.2008.11.006

Banerjee, P., Lenz, D., Robinson, J.P., Rickus, J.L., & Bhunia, A.K. (2008). A novel and simple cell-based detection system with a collagen-encapsulated B-lymphocyte cell line as a biosensor for rapid detection of pathogens and toxins. *Laboratory Investigation; A Journal of Technical Methods and Pathology, 88,* 196-206.

Banik, R.M., Prakash, R., & Upadhyay, S.N. (2008). Microbial biosensor based on whole cell of Pseudomonas sp. for online measurement of p-Nitrophenol. *Sensors and Actuators B: Chemical, 131,* 295-300. http://dx.doi.org/10.1016/j.snb.2007.11.022

Barkay, T., Gillman, M., & Turner, R.R. (1997). Effects of dissolved organic carbon and salinity on bioavailability of mercury. *Applied and Environmental Microbiology, 63,* 4267-4271.

Baroxymeter (2013). Accessed August 31st. http://www.baroxymeter.com/.

Bechor, O., Smulski, D.R, Van Dyk, T.K., LaRossa, R.A, & Belkin, S. (2002). Recombinant microorganisms as environmental biosensors: pollutants detection by Escherichia coli bearing fabA'::lux fusions. *Journal of Biotechnology, 94,* 125-32. http://dx.doi.org/10.1016/S0168-1656(01)00423-0

Belkin, S. (2003). Microbial whole-cell sensing systems of environmental pollutants. *Current Opinion in Microbiology, 6,* 206-212. http://dx.doi.org/10.1016/S1369-5274(03)00059-6

Belkin, S., & Gu, M.B. (2010). *Whole Cell Sensing Systems I: Reporter Cells and Devices* (p.220). Springer, (77-84, 131-154, 155-178).

Ben-Yoav, H., Biran, A., Pedahzur, R, Belkin, S., Buchinger, S., Reifferscheid, G., et al. (2009). A whole cell electrochemical biosensor for water genotoxicity bio-detection. *Electrochimica Acta, 54,* 6113-6118. http://dx.doi.org/10.1016/j.electacta.2009.01.061

Ben-Yoav, H., Elad, T., Shlomovits, O., Belkin, S., & Shacham-Diamand, Y. (2009). Optical modeling of bioluminescence in whole cell biosensors. *Biosensors & Bioelectronics, 24,* 1969-1973. http://dx.doi.org/10.1016/j.bios.2008.10.035

Biran, I., Babai, R., Levcov, K., Rishpon, J., & Ron, E.Z. (2000). Online and in situ monitoring of environmental pollutants: electrochemical biosensing of cadmium. *Environmental Microbiology, 2,* 285-290. http://dx.doi.org/10.1046/j.1462-2920.2000.00103.x

Bond, D.R., & Lovley, D.R. (2003). Electricity production by Geobacter sulfurreducens attached to electrodes. *Applied and Environmental Microbiology, 69,* 1548-1555. http://dx.doi.org/10.1128/AEM.69.3.1548-1555.2003

Borkholder, D. (1998). *Cell based biosensors using microelectrodes.* Stanford University, 27-50.

Braissant, O., Wirz, D., Göpfert, B., & Daniels, A.U. (2010). Use of isothermal microcalorimetry to monitor microbial activities. *FEMS Microbiology Letters, 303,* 1-8. http://dx.doi.org/10.1111/j.1574-6968.2009.01819.x

Bringhurst, R.M., Cardon, Z.G., & Gage, D.J. (2001). Galactosides in the rhizosphere: utilization by Sinorhizobium meliloti and development of a biosensor. *Proceedings of the National Academy of Sciences of the United States of America, 98,* 4540-4545. http://dx.doi.org/10.1073/pnas.071375898

Burmølle, M., Hansen, L.H., Oregaard, G., & Sørensen, S.J. (2003). Presence of N-acyl homoserine lactones in soil detected by a whole-cell biosensor and flow cytometry. *Microbial Ecology, 45,* 226-236. http://dx.doi.org/10.1007/s00248-002-2028-6

Burrell, G.A., & Seibert, F.M. (1916). *Gases found in coal mines.* United States Bureau of Mines, Department of interior. U.S. G.P.O., 23.

Butterworth, F.M., Gunatilaka, A., & Gonsebatt, M.E. (Eds.) (2001). *The "mussel monitor" as biological early warning system.* Boston, MA: Springer US.

Carson, R. (1962). Silent Spring. Boston, MA: Mariner.

Carter, C., Kimbrough, R., Liddle, J., Cline, R., Zack, M., Barthel, W., et al. (1975). Tetrachlorodibenzodioxin: an accidental poisoning episode in horse arenas. *Science, 188,* 738-740. http://dx.doi.org/10.1126/science.1168366

Casavant, N.C., Thompson, D., Beattie, G.A., Phillips, G.J, & Halverson, L.J. (2003). Use of a site-specific recombination-based biosensor for detecting bioavailable toluene and related compounds on roots. *Environmental Microbiology, 5,* 238-249. http://dx.doi.org/10.1046/j.1462-2920.2003.00420.x

Case, A.A., & Coffman, J.R. (1973). Waste oil: toxic for horses. *The Veterinary Clinics of North America, 3,* 273-277.

Ceriotti, L., Kob, A., Drechsler, S., Ponti, J., Thedinga, E., Colpo, P., et al. (2007). Online monitoring of BALB/3T3 metabolism and adhesion with multiparametric chip-based system. *Analytical Biochemistry, 371,* 92-104. http://dx.doi.org/10.1016/j.ab.2007.07.014

Chan, C., Lehmann, M., Chan, K., Chan, P., Chan, C., Gruendig, B., et al. (2000). Designing an amperometric thick-film microbial BOD sensor. *Biosensors and Bioelectronics, 15,* 343-353. http://dx.doi.org/10.1016/S0956-5663(00)00090-7

Chan, C., Lehmann, M., Tag, K., Lung, M., Riedel, K., Gruendig, B., et al. (1999). Measurement of biodegradable substances using the salt-tolerant yeast Arxula adeninivorans for a microbial sensor immobilized with poly(carbamoyl) sulfonate (PCS) part I: construction and characterization of the microbial sensor. *Biosensors and Bioelectronics, 14,* 131-138. http://dx.doi.org/10.1016/S0956-5663(98)00110-9

Chang, I., Moon, H., & Bretschger, O. (2006). Electrochemically active bacteria (EAB) and mediator-less microbial fuel cells. *Journal of Microbiology and Biotechnology, 16,* 163-177.

Chao, H.-P., & Lee, W.-C. (2000). A bioelectrode for penicillin detection based on gluten-membrane-entrapped microbial cells. *Biotechnology and Applied Biochemistry, 32,* 9. http://cx.doi.org/10.1042/BA20000003

Chaudhuri, S.K., & Lovley, D.R. (2003). Electricity generation by direct oxidation of glucose in mediatorless microbial fuel cells. *Nature Biotechnology, 21,* 1229-1232. http://dx.doi.org/10.1038/nbt867

CheckLight Highlighting Water Safety-TOX-SCREEN 3 (2013). Accessed August 30th. http://www.checklight.biz/pcb-tox.php.

Chee, G.-J., Nomura, V, Ikebukuro, K., & Karube, I. (2005). Development of photocatalytic biosensor for the evaluation of biochemical oxygen demand. *Biosensors & Bioelectronics, 21,* 67-73. http://dx.doi.org/10.1016/j.bios.2004.10.005

Chee, G.-J., Nomura, Y., & Karube, I. (1999). Biosensor for the estimation of low biochemical oxygen demand. *Analytica Chimica Acta, 379,* 185-191. http://dx.doi.org/10.1016/S0003-2670(98)00680-1

Chen, D., Cao, Y., Liu, B., & Kong, J. (2002). A BOD biosensor based on a microorganism immobilized on an Al2O3 sol-gel matrix. *Analytical and Bioanalytical Chemistry, 372,* 737-739. http://dx.doi.org/10.1007/s00216-001-1214-6

Chouteau, C., Dzyadevych, S., Durrieu, C., & Chovelon, J.-M. (2005). A bi-enzymatic whole cell conductometric biosensor for heavy metal ions and pesticides detection in water samples. *Biosensors & Bioelectronics, 21,* 273-281. http://dx.doi.org/10.1016/j.bios.2004.09.032

Copeland, R. (2004). *Enzymes: A practical introduction to structure, mechanism, and data analysis.* Wiley-BCH, 11-41.

Curtis, T., Naal, R.M.Z.G., Batt, C., Tabb, J., & Holowka, D. (2008). Development of a mast cell-based biosensor. *Biosensors & Bioelectronics, 23,* 1024-1031. http://dx.doi.org/10.1016/j.bios.2007.10.007

D'Souza, S.F. (2001). Microbial biosensors. *Biosensors & Bioelectronics, 16,* 337-353. http://dx.doi.org/10.1016/S0956-5663(01)00125-7

Daunert, S., Barrett, G., Feliciano, J.S., Shetty, R.S., Shrestha, S., & Smith-Spencer, W. (2000). Genetically engineered whole-cell sensing systems: coupling biological recognition with reporter genes. *Chemical Reviews, 100,* 2705-2738. http://dx.doi.org/10.1021/cr990115p

Dhall, P., Kumar, A., Joshi, A., Saxsena, T.K., Manoharan, A., Makhijani, S.D., et al. (2008). Quick and reliable estimation of BOD load of beverage industrial wastewater by developing BOD biosensor. *Sensors and Actuators B: Chemical, 133,* 478-483. http://dx.doi.org/10.1016/j.snb.2008.03.010

Di Lorenzo, M., Curtis, T.P., Head, I.M., & Scott, K. (2009). A single-chamber microbial fuel cell as a biosensor for wastewaters. *Water research, 43,* 3145-3154. http://dx.doi.org/10.1016/j.watres.2009.01.005

Di Paolantonia, C.L., & Rechnitz, G.A. (1982). Induced bacterial electrode for the potentiometric measurement of tyrosine. *Analytica Chimica Acta, 141,* 1-13. http://dx.doi.org/10.1016/S0003-2670(01)95305-X

Di Paolantonia, C.L., & Rechnitz, G.A. (1983). Stabilized bacteria-based potentiometric electrode for pyruvate. *Analytica Chimica Acta, 148,* 1-12. http://dx.doi.org/10.1016/S0003-2670(00)85146-6

Dubey, R., & Upadhyay, S. (2001). Microbial corrosion monitoring by an amperometric microbial biosensor developed using whole cell of Pseudomonas sp. *Biosensors and Bioelectronics, 16,* 995-1000. http://dx.doi.org/10.1016/S0956-5663(01)00203-2

East, A.K., Mauchline, T.H., & Poole, P.S. (2008). Biosensors for ligand detection. *Advances in Applied Microbiology, 64,* 137-166. http://dx.doi.org/10.1016/S0065-2164(08)00405-X

Elasri, M.O., & Miller, R.V. (1999). Study of the response of a biofilm bacterial community to UV radiation. *Applied and Environmental Microbiology, 65,* 2025-2031.

Emelyanova, E.V., & Reshetilov, A.N. (2002). Rhodococcus erythropolis as the receptor of cell-based sensor for 2,4-dinitrophenol detection: effect of "co-oxidation". *Process Biochemistry, 37,* 683-692. http://dx.doi.org/10.1016/S0032-9592(01)00257-6

Endo, H., Kamata, A., Hoshi, M., Hayashi, T., & Watanabe, E. (1995). Microbial Biosensor System for Rapid Determination of Vitamin B 6. *Journal of Food Science, 60,* 554-557. http://dx.doi.org/10.1111/j.1365-2621.1995.tb09825.x

Environmental technology verification program (2006). Verification Statement Hach Company ToxTrakTM Rapid Toxicity Testing System.

EPA. (2003a). Verification Report Strategic Diagnostics Inc. Deltatox® Rapid Toxicity Testing System.

EPA. (2003b). Verification Statement Interlab Supply, Ltd. Polytox Rapid Toxicity Testing Systems.

EPA. (2005). Technologies and Techniques for Early Warning Systems to Monitor and Evaluate Drinking Water Quality : A State-of-the-Art Review. Office of Research and Development, Washington, 165.

EPA. (2006). Verification Report Checklight Ltd. Toxscreen-II Test Kit.

Farré, M., Pasini, O., Carmen Alonso, M., Castillo, M., & Barceló, D. (2001). Toxicity assessment of organic pollution in wastewaters using a bacterial biosensor. *Analytica Chimica Acta, 426,* 155-165. http://dx.doi.org/10.1016/S0003-2670(00)00826-6

Ferrini, A.M., Mannoni, V., Carpico, G., & Pellegrini, G.E. (2008). Detection and identification of beta-lactam residues in milk using a hybrid biosensor. *Journal of Agricultural and Food Chemistry, 56,* 784-8. http://dx.doi.org/10.1021/jf071479i

Fish Toximeter-Detection of Toxic Substances in the Water-bbe moldaenke (2013). Accessed July 2. http://www.bbe-moldaenke.de/toxicity/daphniatoximeter/.

Fleming, J.T. (2010). Electronic Interfacing with Living Cells. In S. Belkin & M. B. Gu (Eds.). *Whole Cell Sensing Systems I: Reporter Cells and Devices*. Springer. p.155-178.

Fleschin, S., Bala, C., Bunaciu, A.A., Panait, A., & Aboul-Enein, H.Y. (1998). Enalapril microbial biosensor. *Preparative Biochemistry & Biotechnology, 28,* 261-269. http://dx.doi.org/10.1080/10826069808010140

Fluotox (2013). Accessed August 31st. http://www.ifetura.com/es/activites-departements/arnatronic/produits/fluotox.html.

Gäberlein, S., Spener, F., & Zaborosch, C. (2000). Microbial and cytoplasmic membrane-based potentiometric biosensors for direct determination of organophosphorus insecticides. *Applied Microbiology and Biotechnology, 54,* 652-658. http://dx.doi.org/10.1007/s002530000437

Galindo, E., Bautista, D., García, J.L., & Quintero, R. (1990). Microbial sensor for penicillins using a recombinant strain of Escherichia coli. *Enzyme and Microbial Technology, 12,* 642-646. http://dx.doi.org/10.1016/0141-0229(90)90001-7

Giaever, I., & Keese, C.R. (1993). A morphological biosensor for mammalian cells. *Nature,* 366, 591-592. http://dx.doi.org/10.1038/366591a0

Gibbs, J. (1873). A method of geometrical representation of the thermodynamic properties of substances by means of surfaces. *In Transactions of the Connecticut Academy of Arts and Sciences, 2,* 382-404.

Gilchrist, K.H., Giovangrandi, L., Whittington, R.H., & Kovacs, G.T.A. (2005). Sensitivity of cell-based biosensors to environmental variables. *Biosensors & Bioelectronics, 20,* 1397-406. http://dx.doi.org/10.1016/j.bios.2004.06.007

Grieshaber, D., MacKenzie, R , Vörös, J., & Reimhult, E. (2008). Electrochemical biosensors-Sensor principles and architectures. *Sensors, 8,* 1400-1458. http://dx.doi.org/10.3390/s8031400

Gubernot, D.M., Boyer, B.L., & Moses, M.S. (2008). Animals as early detectors of bioevents: veterinary tools and a framework for animal-human integrated zoonotic disease surveillance. *Public Health Reports, 123,* 300-15.

Guedri, H., & Durrieu, C. (2008). A self-assembled monolayers based conductometric algal whole cell biosensor for water monitoring. *Microchimica Acta, 163,* 179-184. http://dx.doi.org/10.1007/s00604-008-0017-2

Guillette Jr, L.J., & Gross, T. (1994). Developmental abnormalities of the gonad and abnormal sex hormone concentrations in juvenile aligators from contaminated and control lakes in Florida. *Environ Health Perspect, 102(3),* 680-688. http://dx.doi.org/10.1289/ehp.94102680

Hach-Lange UK-LUMIStox (2013). Accessed August 31st. http://www.hach-lange.co.uk/view/product/EU-LPV384/LUMIStox?productCode=EU-LPV384.

Han, T.-S., Kim, Y.-C., Sasaki, S., Yano. K., Ikebukuro, K., Kitayama, A., et al. (2001). Microbia sensor for trichloroethylene determination. *Analytica Chimica Acta, 431,* 225-230. http://dx.doi.org/10.1016/S0003-2670(00)01329-5

Han, T.-S., Sasaki, S., Yano, K., Ikebukuro, K, Kitayama, A, Nagamune, T., et al. (2002). Flow injection microbial trichloroethylene sensor. *Talanta, 57,* 271-276. http://dx.doi.org/10.1016/S0039-9140(02)00027-9

Hansen, P., Wittekindt, E., & Sherry, J. (2009). Genotoxicity in the environment (eco-genotoxicity). *Biosensors for Environmental Monitoring of Aquatic Systems, 5,* 203-226. http://dx.doi.org/10.1007/978-3-540-36253-1_8

Haring, C.M., & Meyer, K.F. (1915). Investigation of live-stock conditions and losses in the Selby smoke zone. Bur. Mines, Bull. (U.S.), 98.

Harms, H., Wells, M.C., & van der Meer, J.R. (2006). Whole-cell living biosensors-are they reacy for environmental application? *Applied Microbiology and Biotechnology, 70,* 273-280. http://dx.doi.org/10.1007/s00253-006-0319-4

Haruyama, T. (2006). Cellular biosensing: chemical and genetic approaches. *Analytica Chimica, Acta, 568,* 211-216. http://dx.doi.org/10.1016/j.aca.2005.10.026

Heim, S., Schnieder, I., Binz, D., Vogel, A., & Bilitewski, U. (1999). Development of an automated microbial sensor system. *Biosensors & Bioelectronics, 14,* 187-193. http://dx.doi.org/10.1016/S0956-5663(98)00118-3

Held, M., Schuhmann, W., Jahreis, K., & Schmidt, H.-L. (2002). Microbial biosensor array with transport mutants of Escherichia coli K12 for the simultaneous determination of mono-and disaccharides. *Biosensors & Bioelectronics, 17,* 1089-1094. http://dx.doi.org/10.1016/S0956-5663(02)00103-3

Hernandez, C., Gaviria, L., Segura, S., & Osma, J. (2013). Concept design for a novel confined-bacterial-based biosensor for water quality control. *Pan American Health Care Exchanges (PAHCE), 2013.* http://dx.doi.org/10.1109/PAHCE.2013.6568257

Holm, L.W., Wheat, J.D., Rhode, E.A., & Firch, G. (1953). The treatment of chronic lead poisoning in horses with calcium disodium ethylenediaminetetraacetate. *Journal of the American Veterinary Medical Association, 123,* 383-388.

Holmes, D.E., Bond, D.R., & Lovley, D.R. (2004). Electron Transfer by Desulfobulbus propionicus to Fe(III) and Graphite Electrodes. *Applied and Environmental Microbiology, 70,* 1234-1237. http://dx.doi.org/10.1128/AEM.70.2.1234-1237.2004

Horsburgh, A.M., Mardlin, D.P., Turner, N. L., Henkler, R., Strachan, N., Glover, L.A., et al. (2002). On-line microbial biosensing and fingerprinting of water pollutants. *Biosensors & Bioelectronics, 17,* 495-501. http://dx.doi.org/10.1016/S0956-5663(01)00321-9

Ignatov, O.V., Rogatcheva, S.M., Kozulin, S.V., & Khorkina, N.A. (1997). Acrylamide and acrylic acid determination using respiratory activity of microbial cells. *Biosensors and Bioelectronics, 12,* 105-111. http://dx.doi.org/10.1016/S0956-5663(97)87056-X

Ikebukuro, K., Honda, M., Nakanishi, K., Nomura, Y., Masuda, Y., Yokoyama, K., et al. (1996). Flow-type cyanide sensor using an immobilized microorganism. *Electroanalysis, 8,* 876-879. http://dx.doi.org/10.1002/elan.1140081005

Ikebukuro, K., Miyata, A., Cho, S.J., Nomura, Y., Chang, S.M., Yamauchi, Y., et al. (1996). Microbial cyanide sensor for monitoring river water. *Journal of Biotechnology, 48,* 73-80. http://dx.doi.org/10.1016/0168-1656(96)01399-5

Ikeda, T., & Kano, K. (2001). An electrochemical approach to the studies of biological redox reactions and their applications to biosensors, bioreactors, and biofuel cells. *Journal of bioscience and bioengineering, 92,* 9-18.

Ikeda, T., Kato, K., Maeda. M., Tatsumi, H., Kano, K., & Matsushita, K. (1997). Electrocatalytic properties of Acetobacter aceti cells immobilized on electrodes for the quinone-mediated oxidation of ethanol. *Journal of Electroanalytical Chemistry, 430(1),* 197-204. http://dx.doi.org/10.1016/S0022-0728(97)00164-2

Interagency Coordinating Committee on the Validation of Alternative Methods. (1997). Validation and Regulatory Acceptance of Toxicological Test Methods. National Institute of Environmental Health Sciences.

InterLab Supply-Products-Biological Oxygen Demand (BOD) and Toxicity Testing Technology (2013). Accessed August 30th. http://www.polyseed.com/main/polytox.php.

Jackson, T.F., & Halbert, F.L. (1974). A toxic syndrome associated with the feeding of polybrominated biphenyl-contaminated protein concentrate to dairy cattle. *Journal of the American Veterinary Medical Association, 165,* 437-439.

Jansson, J.K. (2003). Marker and reporter genes: illuminating tools for environmental microbiologists. *Current Opinion in Microbiology, 6,* 310-316. http://dx.doi.org/10.1016/S1369-5274(03)00057-2

Jencks, W.P. (1987). *Catalysis in Chemistry and Enzymology.* Dover Publications, 7-322.

Jha, S K., Kanungo, M., Nath, A., & D'Souza, S.F. (2009). Entrapment of live microbial cells in electropolymerized polyaniline and their use as urea biosensor. *Biosensors & Bioelectronics, 24,* 2637-2642. http://dx.doi.org/10.1016/j.bios.2009.01.024

Jia, J.,Tang, M., Chen, X., Qi, L., & Dong, S. (2003). Co-immobilized microbial biosensor for BOD estimation based on sol-gel derived composite material. *Biosensors & Bioelectronics, 18,* 1023-1029. http://dx.doi.org/10.1016/S0956-5663(02)00225-7

Joyner, D.C., & Lindow, S.E. (2000). Heterogeneity of iron bioavailability on plants assessed with a whole-cell GFP-based bacterial biosensor. *Microbiology (Reading, England), 146(1),* 2435-2445.

Kamei, K., Haruyama, T., Mie, M., Yanagida, Y., Aizawa, M., & Kobatake, E. (2003). Development of immune cellular biosensing system for assessing chemicals on inducible nitric oxide synthase signaling activator. *Analytical Biochemistry, 320,* 75-81. http://dx.doi.org/10.1016/S0003-2697(03)00360-9

Kara, S., Keskinler, B, & Erhan, E. (2009). A novel microbial BOD biosensor developed by the immobilization of P. Syringae in micro-cellular polymers. *Journal of Chemical Technology & Biotechnology, 84,* 511-518. http://dx.doi.org/10.1002/jctb.2071

Karube, I., Wang, Y., Tamiya, E., & Kawarai, M. (1987). Microbial electrode sensor for vitamin B12. *Analytica Chimica Acta, 199,* 93-97. http://dx.doi.org/10.1016/S0003-2670(00)82800-7

Karube, I., Yokoyama, K., Sode, K., & Tamiya, E. (1989). Microbial BOD Sensor Utilizing Thermophilic Bacteria. *Analytical Letters, 22,* 791-801. http://dx.doi.org/10.1080/00032718908051367

Katrlík, J., Švorc, J., Rosenberg, M., & Miertuš, S. (1996). Whole cell amperometric biosensor based on Aspergillus niger for determination of glucose with enhanced upper linearity limit. *Analytica Chimica Acta, 331,* 225-232. http://dx.doi.org/10.1016/0003-2670(96)00209-7

Katrlík, J., Vostiar, I., Sefcovicová, J., Tkác, J., Mastihuba, V., Valach, M., et al. (2007). A novel microbial biosensor based on cells of Gluconobacter oxydans for the selective determination of 1,3-propanediol in the presence of glycerol and its application to bioprocess monitoring. *Analytical and Bioanalytical Chemistry, 388,* 287-295. http://dx.doi.org/10.1007/s00216-007-1211-5

Keenan, P.O., Knight, A.W., Billinton, N., Cahill, P.A., Dalrymple, I.M., Hawkyard, C.J., et al. (2007). Clear and present danger? The use of a yeast biosensor to monitor changes in the toxicity of industrial effluents subjected to oxidative colour removal treatments. *Journal of Environmental Monitoring (JEM), 9,* 1394-1401. http://dx.doi.org/10.1039/b710406e

Kim, G.T., Hyun, M.S., Chang, I.S., Kim, H.J., Park, H.S., Kim, B.H., et al. (2005). Dissimilatory Fe(III) reduction by an electrochemically active lactic acid bacterium phylogenetically related to Enterococcus gallinarum isolated from submerged soil. *Journal of Applied Microbiology, 99,* 978-987. http://dx.doi.org/10.1111/j.1365-2672.2004.02514.x

Kim, H.J., Park, H.S., Hyun, M.S., Chang, I.S., Kim, M., & Kim, B.H. (2002). A mediator-less microbial fuel cell using a metal reducing bacterium, Shewanella putrefaciens. *Enzyme and Microbial Technology, 30,* 145-152. http://dx.doi.org/10.1016/S0141-0229(01)00478-1

Kim, M.-N., & Kwon, H.-S. (1999). Biochemical oxygen demand sensor using Serratia marcescens LSY 4. *Biosensors and Bioelectronics, 14,* 1-7. http://dx.doi.org/10.1016/S0956-5663(98)00107-9

Kirgöz, U.A., Odaci, D., Timur, S., Merkoçi, A., Pazarlioğlu, N., Telefoncu, A., et al. (2006). Graphite epoxy composite electrodes modified with bacterial cells. *Bioelectrochemistry (Amsterdam, Netherlands), 69,* 128-131. http://dx.doi.org/10.1016/j.bioelechem.2005.11.002

Kirgoz, Ü.A., Timur, S., Odaci, D., Pérez, B., Alegret, S., & Merkoçi, A. (2007). Carbon Nanotube Composite as Novel Platform for Microbial Biosensor. *Electroanalysis, 19,* 893-898. http://dx.doi.org/10.1002/elan.200603786

Kitagawa, Y., Ameyama, M., Nakashima, K., Tamiya, E., & Karube, I. (1987). Amperometric alcohol sensor based on an immobilised bacteria cell membrane. *The Analyst, 112,* 1747-1749. http://dx.doi.org/10.1039/an9871201747

Kloss, D., Fischer, M., Rothermel, A., Simon, J.C., & Robitzki, A.A. (2008). Drug testing on 3D in vitro tissues trapped on a microcavity chip. *Lab on a Chip, 8,* 879-884. http://dx.doi.org/10.1039/b800394g

König, A., Reul, T., Harmeling, C., Spener, F., Knoll, M., & Zaborosch, C. (2000). Multimicrobial Sensor Using Microstructured Three-Dimensional Electrodes Based on Silicon Technology. *Analytical Chemistry, 72,* 2022-2028. http://dx.doi.org/10.1021/ac9908391

König, A., Zaborosch, C., & Spener, F. (1997). *Microbial Sensor for Pah in Aqueous Solution Using Solubilizers. Field Screening Europe.* Springer. 203-206. http://dx.doi.org/10.1007/978-94-009-1473-5_47

König, A., Zaborosch, C., Muscat, A., Vorlop, K.-D., & Spener, F. (1996). Microbial sensors for naphthalene using Sphingomonas sp. B1 or Pseudomonas fluorescens WW4. *Applied Microbiology and Biotechnology, 45,* 844-850. http://dx.doi.org/10.1007/s002530050772

Koshland, D.E. (1995). The Key-Lock Theory and the Induced Fit Theory. *Angewandte Chemie International Edition in English,* 33, 2375-2378. http://dx.doi.org/10.1002/anie.199423751

Kramer, K.J.M. (2009). *Continuous Monitoring of Waters by Biological Early Warning Systems. Rapid Chemical and Biological Techniques for Water Monitoring.* John Wiley & Sons, Ltd. 197-219. http://dx.doi.org/10.1002/9780470745427.ch3e

Kumar, S., Kundu, S., Pakshirajan, K, & Dasu, V.V. (2008). Cephalosporins determination with a novel microbial biosensor based on permeabilized Pseudomonas aeruginosa whole cells. *Applied Biochemistry and Biotechnology, 151,* 653-664. http://dx.doi.org/10.1007/s12010-008-8280-6

Kumlanghan, A., Kanatharana, P., Asawatreratanakul, P., Mattiasson, B., & Thavarungkul, P. (2008). Microbial BOD sensor for monitoring treatment of wastewater from a rubber latex industry. *Enzyme and Microbial Technology, 42,* 483-491. http://dx.doi.org/10.1016/j.enzmictec.2008.01.012

Kumlanghan, A., Liu, J., Thavarungkul, P., Kanatharana, P., & Mattiasson, B. (2007). Microbial fuel cell-based biosensor for fast analysis of biodegradable organic matter. *Biosensors & Bioelectronics, 22,* 2939-2944. http://dx.doi.org/10.1016/j.bios.2006.12.014

Kuratsune, M., Yoshimura, T., Matsuzaka, J., &. Yamaguchi, A. (1972). Epidemiologic study on Yusho, a Poisoning Caused by Ingestion of Rice Oil Contaminated with a Commercial Brand of Polychlorinated Biphenyls. *Environmental Health Perspectives, 1,* 119-128.

Kurland, L.T., Faro, S.N., & Siedler, H. (1960). Minamata disease; the outbreak of a neurologic disorder in Minamata, Japan, and its relationship to the ingestion of seafood contaminated by mercury compounds. *World Neurol, 1,* 370-391.

Lechelt, M., Blohm, W., Kirschneit, B., Pfeiffer, M., & Gresens, E. (2000). Monitoring of surface water by ultra-sensitive Daphnia-Toximeter. *Environmental Toxicology, 15,* 390-400. http://dx.doi.org/10.1002/1522-7278(2000)15:5<390::AID-TOX6>3.0.CO;2-H

Lee, J.I., & Karube, I. (1996). Reactor type sensor for cyanide using an immobilized microorganism. *Electroanalysis, 8,* 1117-1120. http://dx.doi.org/10.1002/elan.1140081208

Lee, M.-Y., Kumar, R.A., Sukumaran, S.M., Hogg, M.G., Clark, D.S., & Dordick, J.S. (2008). Three-dimensional cellular microarray for high-throughput toxicology assays. *Proceedings of the National Academy of Sciences of the United States of America*, 105, 59-63. http://dx.doi.org/10.1073/pnas.0708756105

Lehmann, M., Riedel, K., Adler, K., & Kunze, G. (2000). Amperometric measurement of copper ions with a deputy substrate using a novel Saccharomyces cerevisiae sensor. *Biosensors and Bioelectronics, 15,* 211-219. http://dx.doi.org/10.1016/S0956-5663(00)00060-9

Lei, Y., Chen, W., & Mulchandani, A. (2006). Microbial biosensors. *Analytica Chimica Acta, 568,* 200-210. http://dx.doi.org/10.1016/j.aca.2005.11.065

Lei, Y., Mulchandani, P., Chen, W., & Mulchandani, A. (2005). Direct determination of p-nitrophenyl substituent organophosphorus nerve agents using a recombinant Pseudomonas putida JS444-modified Clark oxygen electrode. *Journal of Agricultural and Food Chemistry, 53,* 524-527. http://dx.doi.org/10.1021/jf048943t

Lei, Y., Mulchandani, P., Chen, W., & Mulchandani, A. (2006). Biosensor for Direct Determination of Fenitrothion and EPN Using Recombinant Pseudomonas putida JS444 with Surface Expressed Organophosphorus Hydrolase. 1. Modified Clark Oxygen Electrode. *Sensors, 6,* 466-472. http://dx.doi.org/10.3390/s6040466

Lei, Y., Mulchandani, P., Chen, W., & Mulchandani, A. (2007). Biosensor for direct determination of fenitrothion and EPN using recombinant Pseudomonas putida JS444 with surface-expressed organophosphorous hydrolase. 2. Modified carbon paste electrode. *Applied Biochemistry and Biotechnology, 136,* 243-250. http://dx.doi.org/10.1007/s12010-007-9023-9

Lei, Y., Mulchandani, P., Chen, W., Wang, J, & Mulchandani, A. (2003). A Microbial Biosensor forp-Nitrophenol UsingArthrobacter Sp. *Electroanalysis, 15,* 1160-1164. http://dx.doi.org/10.1002/elan.200390141

Lei, Y., Mulchandani, P., Chen, W., Wang, J, & Mulchandani, A. (2004). Whole cell-enzyme hybrid amperometric biosensor for direct determination of organophosphorous nerve agents with p-nitrophenyl substituent. *Biotechnology and Bioengineering, 85,* 706-713. http://dx.doi.org/10.1002/bit.20022

Lei, Y., Mulchandani, P., Chen, W., Wang, J, & Mulchandani, A. (2005b). Highly Sensitive and Selective Amperometric Microbial Biosensor for Direct Determination of p -Nitropheny-Substituted Organophosphate Nerve Agents. *Environmental Science & Technology, 39,* 8853-8857. http://dx.doi.org/10.1021/es050720b

Leveau, J.H.J., & Lindow, S.E. (2002). Bioreporters in microbial ecology. *Current Opinion in Microbiology, 5,* 259-265. http://dx.doi.org/10.1016/S1369-5274(02)00321-1

Li, F., Tan, T.C., & Lee, Y.K. (1994). Effects of pre-conditioning and microbial composition on the sensing efficacy of a BOD biosensor. *Biosensors and Bioelectronics, 9,* 197-205. http://dx.doi.org/10.1016/0956-5663(94)80122-3

Li, Y.R., & Chu, J. (1991). Study of BOD microbial sensors for waste water treatment control. *Applied Biochemistry and Biotechnology, 28-29,* 855-863. http://dx.doi.org/10.1007/BF02922655

Liu, B., Cui, Y, & Deng J. (1996). Studies on Microbial Biosensor for DL-Phenylalanine and Its Dynamic Response Process. *Analytical Letters, 29,* 1497-1515. http://dx.doi.org/10.1080/00032719608001500

Liu, J., Björnsson, L., & Mattiasson, B. (2000). Immobilised activated sludge based biosensor for biochemical oxygen demand measurement. *Biosensors and Bioelectronics, 14,* 883-893. http://dx.doi.org/10.1016/S0956-5663(99)00064-0

Liu, J., Olsson, G., & Mattiasson, B. (2004a). Short-term BOD (BODst) as a parameter for on-line monitoring of biological treatment process. Part I. A novel design of BOD biosensor for easy renewal of bio-receptor. *Biosensors & Bioelectronics, 20,* 562-570. http://dx.doi.org/10.1016/j.bios.2004.03.008

Liu, J., Olsson, G., & Mattiasson, B. (2004b). Short-term BOD (BODst) as a parameter for on-line monitoring of biological treatment process; Part II: instrumentation of integrated flow injection analysis (FIA) system for BODst estimation. *Biosensors & Bioelectronics, 20,* 571-578. http://dx.doi.org/10.1016/j.bios.2004.03.007

Liu, Q., Cai, H., Xu, Y., Xiao, L., Yang, M., & Wang, P. (2007). Detection of heavy metal toxicity using cardiac cell-based biosensor. *Biosensors & Bioelectronics, 22,* 3224-3229. http://dx.doi.org/10.1016/j.bios.2007.03.005

Marcus, R. (1956a). Electrostatic free energy and other properties of states having nonequilibrium polarization. I. *The Journal of Chemical Physics, 24,* 979-989. http://dx.doi.org/10.1063/1.1742724

Marcus, R. (1956b). On the Theory of Oxidation-Reduction Reactions Involving Electron Transfer. I. *The Journal of Chemical Physics, 24(5),* 966-978. http://dx.doi.org/10.1063/1.1742723

Marty, J.L., Olive, D., & Asano, Y. (1997). Measurement of BOD: Correlation Between 5-Day BOD and Commercial BOD Biosensor Values. *Environmental Technology, 18,* 333-337. http://dx.doi.org/10.1080/09593331808616544

Mascini, M., Memoli, A., & Olana, F. (1989). Microbial sensor for alcohol. *Enzyme and Microbial Technology, 11,* 297-301. http://dx.doi.org/10.1016/0141-0229(89)90045-8

Matsunaga, T., Suzuki, T., & Tomoda, R. (1984). Photomicrobial sensors for selective determination of phosphate. *Enzyme and Microbial Technology, 6,* 355-358. http://dx.doi.org/10.1016/0141-0229(84)90048-6

Mazzei, F., Botrè, F., Lorenti, G., Simonetti, G., Porcelli, F., Scibona, G., et al. (1995). Plant tissue electrode for the determination of atrazine. *Analytica Chimica Acta, 316,* 79-82. http://dx.doi.org/10.1016/0003-2670(95)00343-X

Medtronics Associates. (1970). *Memorandum Report. Horse Pasture Investigation.* Paloalto, CA.

MetPLATETM (2013). Accessed August 31st.
http://www.ees.essie.ufl.edu/homepp/bitton//metplate.asp.

Milner, S.E., & Maguire, A.R. (2012). Recent trends in whole cell and isolated enzymes in enantioselective synthesis. Arkivoc, 321-382. http://dx.doi.org/10.3998/ark.5550190.0013.109

Morgan, E., & Eagleson, K. (1983). Remote biosensing employing fish as real time monitors of water quality events. *Hydrological Applications of Remote Sensing and Remote Data Transmission, 145,* 235-243.

Morgan, E.L.L., & Young, R.C.C. (1984). Automated Multispecies Biosensing System and Development: Advances in Real-Time Water Quality Monitoring. The Biosphere: Problems and Solutions. Proceedings of the *Miami International Symposium on the Biosphere, 25,* 297-301. Elsevier.

Mulchandani, A., Kaneva, I., & Chen, W. (1998). Biosensor for direct determination of organophosphate nerve agents using recombinant Escherichia coli with surface-expressed organophosphorus hydrolase. 2. Fiber-optic microbial biosensor. *Analytical chemistry, 70,* 5042-5046. http://dx.doi.org/10.1021/ac980543l

Mulchandani, A., Mulchandani, P., Kaneva, I., & Chen, W. (1998). Biosensor for Direct Determination of Organophosphate Nerve Agents Using Recombinant Escherichia coli with Surface-Expressed Organophosphorus Hydrolase. 1. Potentiometric Microbial Electrode. *Analytical Chemistry, 70,* 4140-4145. http://dx.doi.org/10.1021/ac9805201

Mulchandani, P., Chen, W., & Mulchandani, A. (2006). Microbial biosensor for direct determination of nitrophenyl-substituted organophosphate nerve agents using genetically engineered Moraxella sp. *Analytica Chimica Acta, 568,* 217-221. http://dx.doi.org/10.1016/j.aca.2005.11.063

Mulchandani, P., Chen, W., Mulchandani, A., Wang, J., & Chen, L. (2001). Amperometric microbial biosensor for direct determination of organophosphate pesticides using recombinant microorganism with surface expressed organophosphorus hydrolase. *Biosensors and Bioelectronics, 16,* 433-437. http://dx.doi.org/10.1016/S0956-5663(01)00157-9

Mulchandani, P., Hangarter, C.M., Lei, Y., Chen, W., & Mulchandani, A. (2005). Amperometric microbial biosensor for p-nitropheno using Moraxella sp.-modified carbon paste electrode. *Biosensors & Bioelectronics, 21,* 523-527. http://dx.doi.org/10.1016/j.bios.2004.11.011

Mulchandani, P., Lei, Y., Chen, W., Wang, J., & Mulchandani, A. (2002). Microbial biosensor for p-nitrophenol using Moraxella sp. *Analytica Chimica Acta, 470,* 79-86. http://dx.doi.org/10.1016/S0003-2670(02)00606-2

Musselmonitor (Mosselmonitor): a biological early warning system (2013). Accessed July 2nd. http://www.mosselmonitor.nl/.

Nakamura, H., Suzuki, K., Ishikuro, H., Kinoshita, S., Koizumi, R., Okuma, S., et al. (2007). A new BOD estimation method employing a double-mediator system by ferricyanide and menadione using the eukaryote Saccharomyces cerevisiae. *Talanta, 72,* 210-216. http://dx.doi.org/10.1016/j.talanta.2006.10.019

Nakanishi, K., Ikebukuro, K., & Karube, I. (1996). Determination of cyanide using a microbial sensor. *Applied Biochemistry and Biotechnology, 60,* 97-106. http://dx.doi.org/10.1007/BF02788064

Nandakumar, R., & Mattiasson, B. (1999a). A low temperature microbial biosensor using immobilised psychrophilic bacteria. *Biotechnology Techniques, 13,* 689-693. http://dx.doi.org/10.1023/A:1008963730069

Nandakumar, R., & Mattiasson, B. (1999b). A microbial biosensor using Pseudomonas putida cells immobilized in an expanded bed reactor for the on-line monitoring of phenolic compounds. *Analytical Letters, 32.*

National Research Council (1983). *Risk assessment in the federal government: Managing the process.* p.191. Washington D.C.: National Academy Press.

National Research Council. (1991). Animals as Sentinels of Environmental Health Hazards (No. PB-91-219576/XAB). National Research Council, Washington, DC (United States). Committee on Animals as Monitors of Environmental Hazards. Chicago.

NCI. (1984). Summary and recommendations: a consensus report. Use of small fish species in carcinogenicity testing. *Natl Cancer Inst Monogr, 65,* 397-404.

Nernst, W., & G. Barr. (1926). *The new heat theorem. p. 78-86.*

Neudörfer, F., & Meyer-Reil, L. (1997). A microbial biosensor for the microscale measurement of bioavailable organic carbon in oxic sediments. Marine ecology progress series. *Oldendorf, 117,* 295-300.

Neufeld, T., Biran, D., Popovtzer, R., Erez, T., Ron, E.Z., & Rishpon, J. (2006). Genetically engineered pfabA pfabR bacteria: an electrochemical whole cell biosensor for detection of water toxicity. *Analytical Chemistry, 78,* 4952-4956. http://dx.doi.org/10.1021/ac052096r

Newman, J., & Thomas-Alyea, K.E. (2012). Electrochemical Systems (Google eBook). John Wiley & Sons. 650

Newman, J.D., & S. J. Setford. (2006). Enzymatic biosensors. *Molecular Biotechnology, 32,* 249-268. http://dx.doi.org/10.1385/MB:32:3:249

Nomura, Y., Ikebukuro, K., Yokoyama, K., Takeuchi, T., Arikawa, Y., Ohno, S., et al. (1994). A Novel Microbial Sensor for Anionic Surfactant Determination. *Analytical Letters, 27,* 3095-3108. http://dx.doi.org/10.1080/00032719408000313

Northeastern Research Center for Wildlife Diseases, Registry of Comparative Pathology, and Institute of Laboratory Animal Resources (U.S.). (1979). *Animals as monitors of environmental pollutants.* National Academy Press. p.218-355.

Odaci, D., Kiralp Kayahan, S., Timur, S., & Toppare, L. (2008). Use of a thiophene-based conducting polymer in microbial biosensing. *Electrochimica Acta, 53,* 4104-4108. http://dx.doi.org/10.1016/j.electacta.2007.12.065

Odaci, D., Sezgintürk, M.K., Timur, S., Pazarlioğlu, N., Pilloton, R., Dinçkaya, E., et al. (2009). Pseudomonas putida based amperometric biosensors for 2,4-D detection. *Preparative Biochemistry & Biotechnology, 39,* 11-19. http://dx.doi.org/10.1080/10826060802589460

Odaci, D., Timur, S., & Telefoncu, A. (2008). Bacterial sensors based on chitosan matrices. *Sensors and Actuators B: Chemical, 134,* 89-94. http://dx.doi.org/10.1016/j.snb.2008.04.013

Odaci, D., Timur, S., & Telefoncu, A. (2009). A microbial biosensor based on bacterial cells immobilized on chitosan matrix. *Bioelectrochemistry (Amsterdam, Netherlands), 75,* 77-82. http://dx.doi.org/10.1016/j.bioelechem.2009.01.002

Ohki, A., Shinohara, K., Ito, O., Naka, K., Maeda, S., Sato, T, et al. (1994). A Bod Sensor Using Klebsiella Oxytoca AS1. *International Journal of Environmental Analytical Chemistry, 56,* 261-269. http://dx.doi.org/10.1080/03067319408034106

Okochi, M., Mima, K., Miyata, M., Shinozaki, Y., Haraguchi, S., Fujisawa, M., et al. (2004). Development of an automated water toxicity biosensor using Thiobacillus ferrooxidans for monitoring cyanides in natural water for a water fltering plant. *Biotechnology and Bioengineering, 87,* 905-911. http://dx.doi.org/10.1002/bit.20193

Pancrazio, J.J., Gray, S.A., Shubin, Y.S., Kulagina, N., Cuttino, D.S., Shaffer, K.M., et al. (2003). A portable microelectrode array recording system incorporating cultured neuronal networks for neurotoxin detection. *Biosensors & Bioelectronics, 18,* 1339-1347. http://dx.doi.org/10.1016/S0956-5663(03)00092-7

Park, H.S., Kim, B.H., Kim, H.S., Kim, H.J., Kim, G.T., Kim, M., et al. (2001). A Novel Electrochemically Active and Fe(III)-reducing Bacterium Phylogenetically Related to Clostridium butyricum Isolated from a Microbial Fuel Cell. *Anaerobe, 7,* 297-306. http://dx.doi.org/10.1006/anae.2001.0399

Parviz, M., & Gross, G.W. (2007). Quantification of zinc toxicity using neuronal networks on microelectrode arrays. *Neurotoxicology, 28,* 520-31. http://dx.doi.org/10.1016/j.neuro.2006.11.006

Peter, J., Hutter, W., Stöllnberger, W., & Hampel, W. (1996). Detection of chlorinated and brominated hydrocarbons by an ion sensitive whole cell biosensor. *Biosensors and Bioelectronics, 11,* 1215-1219. http://dx.doi.org/10.1016/0956-5663(96)88086-9

Pham, C.A., Jung, S.J., Phung, N.T., Lee, J., Chang, I.S., Kim, B.H., et al. (2003). A novel electrochemically active and Fe(III)-reducing bacterium phylogenetically related to Aeromonas hydrophila, isolated from a microbial fuel cell. *FEMS Microbiology Letters, 223,* 129-134. http://dx.doi.org/10.1016/S0378-1097(03)00354-9

Philp, J.C., Balmand, S., Hajto, E., Bailey, M.J., Wiles, S., Whiteley, A.S., et al. (2003). Whole cell immobilised biosensors for toxicity assessment of a wastewater treatment plant treating phenolics-containing waste. *Analytica Chimica Acta, 487,* 61-74. http://dx.doi.org/10.1016/S0003-2670(03)00358-1

Polyak, B., Bassis, E., Novodvorets, A., Belkin, S., & Marks, R. (2000). Optical fiber bioluminescent whole-cell microbial biosensors to genotoxicants. *Water Science & Technology, 42,* 305-311.

Porteous, F., Killham, K., & Meharg, A. (2000). Use of a lux-marked rhizobacterium as a biosensor to assess changes in rhizosphere C flow due to pollutant stress. *Chemosphere, 41,* 1549-1554. http://dx.doi.org/10.1016/S0045-6535(00)00072-2

Pringsheim, P. (1949). *Fluorescence and phosphorescence.* New York: Interscience Publishers, 415-420.

Rabinowitz, P., Gordon, Z., Chudnov, D., Wilcox, M., Odofin, L., Liu, A., et al. (2006). Animals as sentinels of bioterrorism agents. *Emerging infectious diseases, 12,* 647-652 http://dx.doi.org/10.3201/eid1204.051120

Rabinowitz, P., Scotch, M., & Conti, L. (2009). Human and animal sentinels for shared health risks. *Veterinaria italiana, 45,* 23-24.

Rasinger, J.D., Marrazza, G., Briganti, F., Scozzafava, A., Mascini, M., & Turner, A.P.F. (2005). Evaluation of an FIA Operated Amperometric Bacterial Biosensor, Based on Pseudomonas Putida

F1 for the Detection of Benzene, Toluene, Ethylbenzene, and Xylenes (BTEX). *Analytical Letters*, *38,* 1531-1547. http://dx.doi.org/10.1081/AL-200065793

Rasmussen, L., Turner, R., & Barkay, T. (1997). Cell-density-dependent sensitivity of a mer-lux bioassay. *Appl. Envir. Microbiol., 63,* 3291-3293.

Rasmussen, L.D., Sørensen, S.J., Turner, R.R., & Barkay, T. (2000). Application of a mer-lux biosensor for estimating bioavailable mercury in soil. *Soil Biology and Biochemistry, 32,* 639-646. http://dx.doi.org/10.1016/S0038-0717(99)00190-X

Rastogi, S., Kumar, A., Mehra, N., Makhijani, S., Manoharan, A., Gangal, V., et al. (2003). Development and characterization of a novel immobilized microbial membrane for rapid determination of biochemical oxygen demand load in industrial waste-waters. *Biosensors and Bioelectronics, 18,* 23-29. http://dx.doi.org/10.1016/S0956-5663(02)00108-2

Rechnitz, G.A., & Ho, M.Y. (1990). Biosensors based on cell and tissue material. *Journal of Biotechnology, 15,* 201-217. http://dx.doi.org/10.1016/0168-1656(90)90027-9

Renneberg, R., Riedel, K., & Scheller, F. (1985). Microbial sensor for aspartame. *Applied Microbiology and Biotechnology, 21,* 180-181. http://dx.doi.org/10.1007/BF00295116

Reshetilov, A.N., Iliasov, P.V., & Reshetilova, T.A. (2010). *The Microbial Cell Based Biosensors.* In V.S. Somerset (Ed.). *Intelligent and Biosensors* (pp. 289-322). Retrieved from: http://www.intechopen.com/books/intelligent-and-biosensors/the-microbial-cell-based-biosensors

Reshetilov, A.N., Iliasov, P.V., Knackmuss, H.J., & Boronin, A.M. (2000). The Nitrite Oxidizing Activity of Nitrobacter Strains as a Base of Microbial Biosensor for Nitrite *Detection. Analytical Letters, 33,* 29-41. http://dx.doi.org/10.1080/00032710008543034

Reshetilov, A.N., Trotsenko, J.A., Morozova, N.O., Iliasov, P.V., & Ashin, V.V. (2001). Characteristics of Gluconobacter oxydans B-1280 and Pichia methanolica MN4 cell based biosensors for detection of ethanol. *Process Biochemistry, 36,* 1015-1020. http://dx.doi.org/10.1016/S0032-9592(01)00141-8

Rider, T.H., Petrovick, M.S., Nargi, F.E., Harper, J.D., Schwoebel, E.D., Mathews, R.H., et al. (2003). A B cell-based sensor for rapid identification of pathogens. *Science, 301,* 213-215. http://dx.doi.org/10.1126/science.1084920

Riedel, K., Beyersdorf-Radeck, B., Neumann, B., & Schaller, F. (1995). Microbial sensors for determination of aromatics and their chloro derivatives. Part III, Determination of chlorinated phenols using a biosensor containing *Trichosporon beigelii (cutaneum). Applied Microbiology and Biotechnology, 43,* 7-9. http://dx.doi.org/10.1007/BF00170614

Riedel, K., Hensel, J., Rothe, S., Neumann, B., & Scheller, F. (1993). Microbial sensors for determination of aromatics and their chloroderivatives. Part II: Determination of chlorinated phenols using a Rhodococcus-containing biosensor. *Applied Microbiology and Biotechnology, 38.*

Riedel, K., Lehmann, M., Tag, K., Renneberg, R. & Kunze, G. (1998). Arxula adeninivorans Based Sensor for the Estimation of BOD. *Analytical Letters, 31,* 1-12. http://dx.doi.org/10.1080/00032719808001829

Riedel, K., Naumov, A.V., Boronin, A.M., Golovleva, L.A., Stein, H.J., & Scheller, F. (1991). Microbial sensors for determination of aromatics and their chloroderivatives I. Determination of

3-chlorobenzoate using a Pseudomonas-containing biosensor. *Applied Microbiology and Biotechnology, 35.*

Riedel, K., Renneberg, R., Kühn, M., & Scheller, F. (1988). A fast estimation of biochemical oxygen demand using microbial sensors. *Applied Microbiology and Biotechnology*, 28. http://cx.doi.org/10.1007/BF00250463

Roda, A., Pasini, P., Mirasoli, M., Guardigli, M., Russo, C., Musiani, M., et al. (2001). Sensitive determination of urinary mercury(II) by a bioluminescent transgenic bacteria-based biosensor. *Analytical Letters, 34,* 29-41. http://dx.doi.org/10.1081/AL-100002702

Rosen, R., Davidov, Y., LaRossa, R.A., & Belkin, S. (2000). Microbial sensors of ultraviolet radiation based on recA':lux fusions. *Applied biochemistry and biotechnology, 89,* 151-160. http://dx.doi.org/10.1385/ABAB:89:2-3:151

Rotariu, L., & Bala, C. (2003). New Type of Ethanol Microbial Biosensor Based on a Highly Sensitive Amperometric Oxygen Electrode and Yeast Cells. *Analytical Letters, 36,* 2459-2471. http://dx.doi.org/10.1081/AL-120024335

Rotariu, L., Bala, C., & Magearu, V. (2000). Use of yeast cells for selective determination of sucrose. *Revue Roumaine de Chimie, 45,* 21-26.

Rotariu, L., Bala, C., & Magearu, V. (2002). Yeast cells sucrose biosensor based on a potentiometric oxygen electrode. *Analytica Chimica Acta, 458,* 215-222. http://dx.doi.org/10.1016/S0003-2670(01)01529-X

Rotariu, L., Bala, C., & Magearu, V. (2004). New potentiometric microbial biosensor for ethanol determination in alcoholic beverages. *Analytica Chimica Acta, 513,* 119-123. http://dx.doi.org/10.1016/j.aca.2003.12.048

Rouillon, R., Sole, M., Carpentier, R., & Marty, J.-L. (1995). Immobilization of thylakoids in polyvinylalcohol for the detection of herbicides. *Sensors and Actuators B: Chemical, 27,* 477-479. http://dx.doi.org/10.1016/0925-4005(94)01645-X

Rouillon, R., Tocabens, M., & Carpentier, R. (1999). A photoelectrochemical cell for detecting pollutant-induced effects on the activity of immobilized cyanobacterium Synechococcus sp. PCC 7942. *Enzyme and Microbial Technology, 25,* 230-235. http://dx.doi.org/10.1016/S0141-0229(99)00033-2

Rubner, M. (1911). Die Kalorimetrie. In *Handbuch der physiologischen Methodik. Bd. 1,* 150-228.

Rudolph, A.S., & Reasor, J. (2001). Cell and tissue based technologies for environmental detection and medical diagnostics. *Biosensors & Bioelectronics, 16,* 429-431. http://dx.doi.org/10.1016/S0956-5663(01)00156-7

Safronova, O., Khichenko, V., & Shtark, M. (1995). Possible clinical applications of tissue and cell biosensors. *Biomedical Engineering, 29,* 39-46. http://dx.doi.org/10.1007/BF00558879

Salis, H., Tamsir, A., & Voigt, C. (2009). Engineering bacterial signals and sensors. *Contributions to Microbiology, 16,* 194-225. http://dx.doi.org/10.1159/000219381

Sargeetha, S., Sugandhi, G., Murugesan, M., Murali Madhav, V., Berchmans, S., Rajasekar, R., et al. (1996). Torulopsis candida based sensor for the estimation of biochemical oxygen demand and its evaluation. *Electroanalysis, 8,* 698-701. http://dx.doi.org/10.1002/elan.1140080718

Schmidt, A., Standfuß-Gabisch, C., & Bilitewski, U. (1996). Microbial biosensor for free fatty acids using an oxygen electrode based on thick film technology. *Biosensors and Bioelectronics, 11,* 1139-1145. http://dx.doi.org/10.1016/0956-5663(96)82336-0

Schreiter, P.P., Gillor, O., Post, A., Belkin, S., Schmid, R.D., & Bachmann, T.T. (2001). Monitoring of phosphorus bioavailability in water by an immobilized luminescent cyanobacterial reporter strain. *Biosensors & Bioelectronics, 16,* 811-818. http://dx.doi.org/10.1016/S0956-5663(01)00224-X

Schwabe, C.W. (1984). *Animal monitors of the environment. Veterinary Medicine and Human Health* (3rd ed.). Lippincott Williams & Wilkins, Baltimore. 562-578

Seki A., Kawakubo, K., Iga, M., & Nomura, S. (2003). Microbial assay for tryptophan using silicon-based transducer. *Sensors and Actuators B: Chemical, 94,* 4. http://dx.doi.org/10.1016/S0925-4005(03)00381-2

Selifonova, O., Burlage, R., & Barkay, T. (1993). Bioluminescent sensors for detection of bioavailable Hg(II) in the environment. *Applied and Environmental Microbiology, 59,* 3083-3090.

Shank, E.A., & Kolter, R. (2009). New developments in microbial interspecies signaling. *Current Opinion in Microbiology, 12,* 205-214. http://dx.doi.org/10.1016/j.mib.2009.01.003

Shimomura-Shimizu, M., & Karube, I. (2010). Applications of Microbial Cell Sensors. In Whole cell sensing system II (pp. 1-30). Springer Berlin Heidelberg. Chicago.

Shitanda, I., Takamatsu, S., Watanabe, K., & Itagaki, M. (2009). Amperometric screen-printed algal biosensor with fbw injection analysis system for detection of environmental toxic compounds. *Electrochimica Acta, 54,* 4933-4936. http://dx.doi.org/10.1016/j.electacta.2009.04.005

Siegfried, E. (2011). Genes Code for Proteins. In Siegfried, E. (Ed.). *Lewin's Genes X, 136,* 26-41. Jones and Bartlett Publishers.

Slaughter, G.E., & Hobson, R. (2009). An impedimetric biosensor based on PC 12 cells for the monitoring of exogenous agents. *Biosensors & Bioelectronics, 24,* 1153-1158. http://dx.doi.org/10.1016/j.bios.2008.06.060

Smutok, O., Dmytruk, K., Gonchar, M., Sibirny, A., & Schuhmann, W. (2007). Permeabilized cells of flavocytochrome b2 over-producing recombinant yeast Hansenula polymorpha as biological recognition element in amperometric lactate biosensors. *Biosensors & Bioelectronics, 23,* 599-605. http://dx.doi.org/10.1016/j.bios.2007.06.021

Sørensen, S.J., Burmølle, M., & Hansen, L.H. (2006). Making bio-sense of toxicity: new developments in whole-cell biosensors. *Current Opinion in Biotechnology, 17,* 11-16. http://dx.doi.org/10.1016/j.copbio.2005.12.007

Stein, N.E., Keesman, K.J., Hamelers, H.V.M., & van Straten, G. (2011). Kinetic models for detection of toxicity in a microbial fuel cell based biosensor. *Biosensors & Bioelectronics, 26,* 3115-3120. http://dx.doi.org/10.1016/j.bios.2010.11.049

Stiffey, A.V., & Nicolaids, T.G. (1995). U.S. Patent No. 5,580,785. Washington, D.C.: U.S. Patent and Trademark Office. Chicago. Issued December 3, 1996.

Stockman, S. (1916). Cases of poisoning in cattle by feeding on a meal from soybean after extraction of the oil. *Journal of Comparative Pathology and Therapeutics, 29,* 95-107. http://dx.doi.org/10.1016/S0368-1742(16)80007-7

Storey, M.V., van der Gaag, B., & Burns, B.P. (2011). Advances in on-line drinking water quality monitoring and early warning systems. *Water Research, 45,* 741-747. http://dx.doi.org/10.1016/j.watres.2010.08.049

Stoytcheva, M., Zlatev, R., Valdez, B., Magnin, J.-P., & Velkova, Z. (2006). Electrochemical sensor based on Arthrobacter globiformis for cholinesterase activity determination. *Biosensors & Bioelectronics, 22,* 1-9. http://dx.doi.org/10.1016/j.bios.2005.11.013

Su, L., Jia, W., Hou, C., & Lei, Y. (2011). Microbial biosensors: a review. *Biosensors & Bioelectronics, 26,* 1788-1799. http://dx.doi.org/10.1016/j.bios.2010.09.005

Subrahmanyam, S., Kodandapani, N., Shanmugam, K., Moovarkumuthalvan, K., Jeyakumar, D., & Subramanian, T.V. (2001). Development of a Sensor for Acetic Acid Based onFusarium solani. *Electroanalysis, 13,* 1275-1278. http://dx.doi.org/10.1002/1521-4109(200110)13:15<1275::AID-ELAN1275>3.0.CO;2-W

Subrahmanyam, S., Shanmugam, K., Subramanian, T.V., Murugesan, M., Madhav, V.M., & Jeyakumar, D. (2001). Development of Electrochemical Microbial Biosensor for Ethanol Based onAspergillus niger. *Electroanalysis, 13,* 944-948. http://dx.doi.org/10.1002/1521-4109(200107)13:11<944::AID-ELAN944>3.0.CO;2-D

Sumathi, R., Rajasekar, R., & Narasimham, K.C. (2000). Acetobacter peroxydans based electrochemical biosensor for hydrogen peroxide. *Bulletin of Electrochemistry*. Central Electrochemical Research Institute.

Suzuki, H., Tamiya, E., & Karube, I. (1987). An amperometric sensor for carbon dioxide based on immobilized bacteria utilizing carbon dioxide. *Analytica Chimica Acta, 199,* 85-91. http://dx.doi.org/10.1016/S0003-2670(00)82799-3

Svitel, J., Curilla, O., & Tkác, J. (1998). Microbial cell-based biosensor for sensing glucose, sucrose or lactose. *Biotechnology and applied biochemistry, 27(2),* 153-158.

SYSTEA S.P.A. (2013). Accessed August 30. http://www.systea.it/.

Tag, K., Kwong, A.W., Lehmann, M., Chan, C., Renneberg, R., Riedel, K., et al. (2000). Fast detection of high molecular weight substances in wastewater based on an enzymatic hydrolysis combined with theArxula BOD sensor system. *Journal of Chemical Technology & Biotechnology, 75,* 1080-1082. http://dx.doi.org/10.1002/1097-4660(200011)75:11<1080::AID-JCTB311>3.0.CO;2-#

Tag, K., Lehmann, M., Chan, C., Renneberg, R., Riedel, K., & Kunze, G. (1998). Arxula adeninivoransLS3 as suitable biosensor for measurements of biodegradable substances in salt water. *Journal of Chemical Technology & Biotechnology, 73,* 385-388. http://dx.doi.org/10.1002/(SICI)1097-4660(199812)73:4<385::AID-JCTB975>3.0.CO;2-B

Tag, K., Lehmann, M., Chan, C., Renneberg, R., Riedel, K., & Kunze, G. (2000) Measurement of biodegradable substances with a mycelia-sensor based on the salt tolerant yeast Arxula adeninivorans LS3. *Sensors and Actuators B: Chemical, 67,* 142-148. http://dx.doi.org/10.1016/S0925-4005(00)00404-4

Tag, K., Riedel, K., Bauer, H.-J., Hanke, G., Baronian, K.H.R., & Kunze, G. (2007). Amperometric detection of Cu2+ by yeast biosensors using flow injection analysis (FIA). *Sensors and Actuators B: Chemical, 122,* 403-409. http://dx.doi.org/10.1016/j.snb.2006.06.007

Tan, T.C., & Qian, Z. (1997). Dead Bacillus subtilis cells for sensing biochemical oxygen demand of waters and wastewaters. *Sensors and Actuators B: Chemical, 40,* 65-70. http://dx.doi.org/10.1016/S0925-4005(97)00013-0

Tan, T.C., & Wu, C. (1999). BOD sensors using multi-species living or thermally killed cells of a BODSEED microbial culture. *Sensors and Actuators B: Chemical, 54,* 252-260. http://dx.doi.org/10.1016/S0925-4005(99)00113-6

Tan, T.C., Li, F., & Neoh, K.G. (1993). Measurement of BOD by initial rate of response of a microbial sensor. *Sensors and Actuators B: Chemical, 10,* 137-142. http://dx.doi.org/10.1016/0925-4005(93)80037-C

Tan, T.C., Li, F., Neoh, K.G., & Lee, Y.K. (1992). Microbial membrane-modified dissolved oxygen probe for rapid biochemical oxygen demand measurement. *Sensors and Actuators B: Chemical, 8,* 167-172. http://dx.doi.org/10.1016/0925-4005(92)80175-W

Taranova, L., Semenchuk, I., Manolov, T., Iliasov, P., & Reshetilov, A. (2002). Bacteria-degraders as the base of an amperometric biosensor for detection of anionic surfactants. *Biosensors and Bioelectronics, 17,* 635-640. http://dx.doi.org/10.1016/S0956-5663(01)00307-4

Taranova, L.A., Fesay, A.P., Ivashchenko, G.V., Reshetilov, A.N., Winther-Nielsen, M., & Emneus, J. (2004). Comamonas testosteroni Strain TI as a Potential Base for a Microbial Sensor Detecting Surfactants. *Applied Biochemistry and Microbiology, 40,* 404-408. http://dx.doi.org/10.1023/B:ABIM.0000033919.64525.5a

Tauber, M., Rosen, R., & Belkin, S. (2001). Whole-cell biodetection of halogenated organic acids. *Talanta, 55,* 959-964. http://dx.doi.org/10.1016/S0039-9140(01)00492-1

Tecon, R., & van der Meer, J.R. (2008). Bacterial Biosensors for Measuring Availability of Environmental Pollutants. *Sensors, 8,* 4062-4080. http://dx.doi.org/10.3390/s8074062

Thévenot, D., Toth, K., Durst, R., & Wilson, G. (2001). Electrochemical biosensors: recommended definitions and classification. *Biosensors and Bioelectronics, 16,* 121-131. http://dx.doi.org/10.1016/S0956-5663(01)00115-4

Thomas, C. (2013). Genetic manipulation of bacteria. XXX *iii*. Accessed August 30th, 2013. *www.eolss.net.*

Thouand, G., Horry, H., Durand, M.J., Picart, P., Bendriaa, L., Daniel, P., et al. (2003). Development of a biosensor for on-line detection of tributyltin with a recombinant bioluminescent Escherichia coli strain. *Applied microbiology and biotechnology, 62,* 218-225. http://dx.doi.org/10.1007/s00253-003-1279-6

Tibazarwa, C., Corbisier, P., Mench, M., Bossus, A., Solda, P., Mergeay, M., et al. (2001). A microbial biosensor to predict bioavailable nickel in soil and its transfer to plants. *Environmental Pollution (Barking, Essex), 113,* 19-26. http://dx.doi.org/10.1016/S0269-7491(00)00177-9

Tiensing, T., Strachan, N., & Paton, G.I. (2002). Evaluation of interactive toxicity of chlorophenols in water and soil using lux-marked biosensors. *Journal of Environmental Monitoring (JEM), 4,* 482-489. http://dx.doi.org/10.1039/b202070j

Timur, S., Anik, U., Odaci, D., & Gorton, L. (2007). Development of a microbial biosensor based on carbon nanotube (CNT) modified electrodes. *Electrochemistry Communications, 9,* 1810-1815. http://dx.doi.org/10.1016/j.elecom.2007.04.012

Timur, S., Della Seta, L., Pazarlioğlu, N., Pilloton, R., & Telefoncu, A. (2004). Screen printed graphite biosensors based on bacterial cells. *Process Biochemistry, 39,* 1325-1329. http://dx.doi.org/10.1016/S0032-9592(03)00265-6

Timur, S., Haghighi, B., Tkac, J., Pazarlioğlu, N., Telefoncu, A., & Gorton, L. (2007). Electrical wiring of Pseudomonas putida and Pseudomonas fluorescens with osmium redox polymers. *Bioelectrochemistry (Amsterdam, Netherlands), 71,* 38-45. http://cx.doi.org/10.1016/j.bioelechem.2006.08.001

Timur, S., Pazarlioğlu, N., Pilloton, R., & Telefoncu, A. (2003). Detection of phenolic compounds by thick film sensors based on Pseudomonas putida. *Talanta, 61,* 87-93. http://dx.doi.org/10.1016/S0039-9140(03)00237-6

Tkáč, J., Gemeiner, P., Švitel, J., Benikovský, T., Šturdík, E., Vala, V, et al. (2000). Determination of total sugars in lignocellulose hydrolysate by a mediated Gluconobacter oxydans biosensor. *Analytica Chimica Acta, 420,* 1-7. http://dx.doi.org/10.1016/S0003-2670(00)01001-1

Tkáč, J., Voštiar, I., Gemeiner, P., & Šturdík, E. (2002). Stabilization of ferrocene leakage by physical retention in a cellulose acetate membrane. The fructose biosensor. *Bioelectrochemistry, 55,* 149-151. http://dx.doi.org/10.1016/S1567-5394(01)00130-X

Tkac, J., Vostiar, I., Gorton, L., Gemeiner, P., & Sturdik, E. (2003). Improved selectivity of microbial biosensor using membrane coating. Application to the analysis of ethanol during fermentation. *Biosensors and Bioelectronics, 18,* 1125-1134. http://dx.doi.org/10.1016/S0956-5663(02)00244-0

Togo, C.A., Wutor, V.C., Limson, J.L., & Pletschke, B.I. (2007). Novel detection of Escherichia coli beta-D-glucuronidase activity using a microbially-modified glassy carbon electrode and its potential for faecal pollution monitoring. *Biotechnology Letters, 29,* 531-537. http://dx.doi.org/10.1007/s10529-006-9282-5

Tom-Petersen, A., Hosbond, C., & Nybroe, O. (2001). Identification of copper-induced genes in Pseudomonas fluorescens and use of a reporter strain to monitor bioavailable copper in soil. *FEMS Microbiology Ecology, 38,* 59-67. http://dx.doi.org/10.1111/j.1574-6941.2001.tb00882.x

Tong, C., Shi, B., Xiao, X., Liao, H., Zheng, Y., Shen, G., et al. (2009). An Annexin V-based biosensor for quantitatively detecting early apoptotic cells. *Biosensors & Bioelectronics, 24,* 1777-1782. http://dx.doi.org/10.1016/j.bios.2008.07.040

ToxTrak™ Toxicity Reagent Set, 25-49 tests-Overview | Hach (2013). Accessed August 30th. http://www.hach.com/toxtrak-toxicity-reagent-set-25-49-tests/product?id=7640273469.

Trask, O.J., Baker, A., Williams, R.G., Nickischer, D., Kandasamy, R., Laethem, C., et al. (2006). Assay development and case history of a 32K-biased library high-content MK2-EGFP translocation screen to identify p38 mitogen-activated protein kinase inhibitors on the ArrayScan 3.1 imaging platform. *Methods in Enzymology, 414,* 419-439. http://dx.doi.org/10.1016/S0076-6879(06)14023-9

Tront, J.M., Fortner, J.D., Plötze, M., Hughes, J.B., & Puzrin, A.M. (2008). Microbial fuel cell biosensor for in situ assessment of microbial activity. *Biosensors & Bioelectronics, 24,* 586-590. http://dx.doi.org/10.1016/j.bios.2008.06.006

Trosok, S.P., Driscoll, B.T., & Luong, J.H.T. (2001). Mediated microbial biosensor using a novel yeast strain for wastewater BOD measurement. *Applied Microbiology and Biotechnology, 56,* 550-554. http://dx.doi.org/10.1007/s002530100674

Tuncagil, S., Odaci, D., Varis, S., Timur, S., & Toppare, L. (2009). Electrochemical polymerization of 1-(4-nitrophenyl)-2,5-di(2-thienyl)-1 H-pyrrole as a novel immobilization platform for microbial sensing. *Bioelectrochemistry (Amsterdam, Netherlands), 76,* 169-174. http://dx.doi.org/10.1016/j.bioelechem.2009.05.001

Tuncagil, S., Odaci, D., Yildiz, E., Timur, S., & Toppare, L. (2009). Design of a microbial sensor using conducting polymer of 4-(2,5-di(thiophen-2-yl)-1H-pyrrole-1-l) benzenamine. *Sensors and Actuators B: Chemical, 137,* 42-47. http://dx.doi.org/10.1016/j.snb.2008.10.067

U.S. Congress Act. Public Law 103-43. 42 U.S.C (1993). Government Printing Office, U.S.

Ukeda, H., Wagner, G., Bilitewski, U., & Schmid, R.D. (1992). Flow injection analysis of short-chain fatty acids in milk based on a microbial electrode. *Journal of Agricultural and Food Chemistry, 40,* 2324-2327. http://dx.doi.org/10.1021/jf00023a053

Ukeda, H., Wagner, G., Weis, G., Miller, M., Klostermeyer, H., & Schmid, R.D. (1992). Application of a microbial sensor for determination of short-chain fatty acids in raw milk samples. *Zeitschrift für Lebensmittel-Untersuchung und -Forschung, 195,* 1-2. http://dx.doi.org/10.1007/BF01197829

Ulstrup, J., & Jortner, J. (1975). The effect of intramolecular quantum modes on free energy relationships for electron transfer reactions. *The Journal of Chemical Physics, 63,* 4358. http://dx.doi.org/10.1063/1.431152

Upton, J., & Pickin, S.R. (1996). Amtox [TM]-A New Concept for Rapid Nitrification Inhibition Testing Applicable to the Laboratory and On-line at Treatment Works. *Special Publications of the Royal Society of Chemistry, 193,* 54-63.

Valach, M., Katrlík, J., Šturdík, E., & Gemeiner, P. (2009). Ethanol Gluconobacter biosensor designed for fbw injection analysis. *Sensors and Actuators B: Chemical, 138,* 581-586. http://dx.doi.org/10.1016/j.snb.2009.02.017

Van Der Meer, J.R. (2011). *Bacterial Sensors: Synthetic Design and Application Principles* (Google eBook). Morgan & Claypool Publishers. p. 153.

Van der Meer, J.R., & Belkin, S. (2010). Where microbiology meets microengineering: design and applications of reporter bacteria. *Nature reviews. Microbiology, 8,* 511-522. http://dx.doi.org/10.1038/nrmicro2392

Van der Schalie, W.H. (1977). *The Utilization of Aquatic Organisms for Continuous and Automatic Monitoring of the Toxicity of Industrial Waste Effluents.* Virginia Polytechnic Institute and State University. p.334.

Van der Schalie, W.H., Gardner, H.S., Bantle, J.A., De Rosa, C.T., Finch, R.A., Reif, J.S., et al. (1999). Animals as sentinels of human health hazards of environmental chemicals. *Environmental Health Perspectives, 107,* 309-315. http://dx.doi.org/10.1289/ehp.99107309

Van der Schalie, W.H., James, R.R., & Gargan, T.P. (2006). Selection of a battery of rapid toxicity sensors for drinking water evaluation. *Biosensors & Bioelectronics, 22,* 18-27. http://dx.doi.org/10.1016/j.bios.2005.11.019

Van der Schalie, W.H., Shedd, T.R., Knechtges, P.L., & Widder, M.W. (2001). Using higher organisms in biological early warning systems for real-time toxicity detection. *Biosensors & Bioelectronics, 16,* 457-465. http://dx.doi.org/10.1016/S0956-5663(01)00160-9

Van Kampen, K.R., James, L.F., Rasmussen, J., Huffaker, R.H., & Fawcett, M.O. (1969). Organic phosphate poisoning of sheep in Skull Valley, Utah. *Journal of the American Veterinary Medical Association, 154,* 623-630.

Verma, N., & Singh, M. (2003). A disposable microbial based biosensor for quality control in milk. *Biosensors and Bioelectronics, 18,* 1219-1224. http://dx.doi.org/10.1016/S0956-5663(03)00085-X

Veterinarian. (1874a). The effects of fog on cattle in London. *Veterinarian, 47,* 1-4.

Veterinarian. (1874b). The effects of the recent fog on the Smithfield Show and the London dairies. *Veterinarian, 47,* 32-33.

Voronova, E.A., Iliasov, P.V., & Reshetilov, A.N. (2008). Development, Investigation of Parameters and Estimation of Possibility of Adaptation of Pichia angusta Based Microbial Sensor for Ethanol Detection. *Analytical Letters, 41,* 377-391. http://dx.doi.org/10.1080/00032710701645729

Walther, J.D., & Wurster, L. (2007). Ecbc-tr-517 environmental sentinel biomonitor (esb) system technology assessment. p. 17. Edgewood. Chemical Biological Center.

Waters, C.M., & Bassler, B.L. (2005). Quorum sensing: cell-to-cell communication in bacteria. *Annual Review of Cell and Developmental Biology, 21,* 319-346. http://dx.doi.org/10.1146/annurev.cellbio.21.012704.131001

Welborn, J., Allen, R., Byker, G., DeGrow, S., Hertel, J., Noordhoek, R., et al. (1975). *The Contamination Crisis in Michigan: Polybrominated Biphenyls.* Lansing, MI.

Wells, M., Gösch, M., Rigler, R., Harms, H., Lasser, T., & van der Meer, J.R. (2005). Ultrasensitive reporter protein detection in genetically engineered bacteria. *Analytical Chemistry, 77,* 2683-2689. http://dx.doi.org/10.1021/ac048127k

Wen, G., Zheng, J., Zhao, C., Shuang, S., Dong, C., & Choi, M.M.F. (2008). A microbial biosensing system for monitoring methane. *Enzyme and Microbial Technology, 43,* 257-261. http://dx.doi.org/10.1016/j.enzmictec.2008.04.006

Werlen, C., Jaspers, M.C.M., & van der Meer, J.R. (2004). Measurement of biologically available napthalene in gas and aqueous phases by use of a Pseudomonas putida biosensor. *Applied and Environmental Microbiology, 70,* 43-51. http://dx.doi.org/10.1128/AEM.70.1.43-51.2004

Wijesuriya, D.C., & Rechnitz, G.A. (1993). Biosensors based on plant and animal tissues. *Biosensors and Bioelectronics, 8,* 155-160. http://dx.doi.org/10.1016/0956-5663(93)85027-L

Wilson, G.S., & Hu, Y. (2000). Enzyme-based biosensors for in vivo measurements. *Chemical Reviews, 100,* 2693-2704. http://dx.doi.org/10.1021/cr990003y

Xu, G., Ye, X., Qin, L., Xu, Y., Li, Y., Li, R., et al. (2005). Cell-based biosensors based on light-addressable potentiometric sensors for single cell monitoring. *Biosensors & Bioelectronics, 20,* 1757-1563. http://dx.doi.org/10.1016/j.bios.2004.06.037

Yagi, K. (2007). Applications of whole-cell bacterial sensors in biotechnology and environmental science. *Applied Microbiology and Biotechnology, 73,* 1251-1258. http://dx.doi.org/10.1007/s00253-006-0718-6

Yang, Z., Sasaki, S., Karube, I., & Suzuki, H. (1997). Fabrication of oxygen electrode arrays and their incorporation into sensors for measuring biochemical oxygen demand. *Analytica Chimica Acta, 357,* 41-49. http://dx.doi.org/10.1016/S0003-2670(97)00560-6

Yang, Z., Suzuki, H., Sasaki, S., & Karube, I. (1996). Disposable sensor for biochemical oxygen demand. *Applied Microbiology and Biotechnology, 46,* 10-14. http://dx.doi.org/10.1016/S0003-2670(97)00560-6

Yeni, F., Odaci, D., & Timur, S. (2008). Use of Eggshell Membrane as an Immobilization Platform in Microbial Sensing. *Analytical Letters, 41,* 2743-2758. http://dx.doi.org/10.1080/00032710802363594

Yoshida, N., Hoashi, J., Morita, T., McNiven, S.J., Nakamura, H., & Karube, I. (2001). Improvement of a mediator-type biochemical oxygen demand sensor for on-site measurement. *Journal of Biotechnology, 88,* 269-275. http://dx.doi.org/10.1016/S0168-1656(01)00282-6

Yoshida, N., Yano, K., Morita, T., McNiven, S.J., Nakamura, H., & Karube, I. (2000). A mediator-type biosensor as a new approach to biochemical oxygen demand estimation. *The Analyst, 125,* 2280-2284. http://dx.doi.org/10.1016/S0168-1656(01)00282-6

Zlatev, R., Magnin, J.-P., Ozil, P., & Stoytcheva, M. (2006a). Bacterial sensors based on Acidithiobacillus ferrooxidans Part II. Cr(VI) determination. *Biosensors & Bioelectronics, 21,* 1501-1506. http://dx.doi.org/10.1016/j.bios.2005.07.004

Zlatev, R., Magnin, J.-P., Ozil, P., & Stoytcheva, M. (2006b). Bacterial sensors based on Acidithiobacillus ferrooxidans Part I. Fe2+ and S2O32- determination. *Biosensors & Bioelectronics, 21,* 1493-1500. http://dx.doi.org/10.1016/j.bios.2005.07.007

Zlatev, R., Magnin, J.-P., Ozil, P., & Stoytcheva, M. (2006c). Acidithiobacillus ferrooxidans fixation on mercuric surfaces and its application in stripping voltammetry. *Biosensors & Bioelectronics, 21,* 1753-1759. http://dx.doi.org/10.1016/j.bios.2005.09.001

Chapter 4

Electrochemical biosensors for the detection of microcystins: Recent advances and perspectives

Audrey Sassolas, Akhtar Hayat, Jean-Louis Marty

IMAGES EA 4218, bât. S, University of Perpignan, 52 avenue Paul Alduy, 66860 Perpignan Cedex, France.

audrey.sassolas@gmail.com, akhtarloona@gmail.com, jlmarty@univ-perp.fr

Doi: http://dx.doi.org/10.3926/oms.123

Referencing this chapter

Sassolas, A., Hayat, A., Marty, J.L. (2014). Electrochemical biosensors for the detection of microcystins: Recent advances and perspectives. In M. Stoytcheva & J.F. Osma (Eds.). *Biosensors: Recent Advances and Mathematical Challenges.* Barcelona: España, OmniaScience. pp. 97-109.

1. Introduction

Water blooms of toxic cyanobacteria (blue-green algae) represent a serious problem because of the potent toxins that can be released by these algae (Dawson, 1998). Toxin-producing microalgae species have a negative influence on the environment, food safety and health (de Figueiredo et al., 2004). Microcystins (MCs) are a group of cyanobacterial toxins that are mainly produced by microcystis, which appear in lake, ponds, reservoirs and rivers with low turbidity flow regimes. More than 80 structural variants of MCs are known, and each one shows very different toxicity levels. These toxins are cyclic heptapeptides with the general structure cyclo-(d-Ala-X-d-MeAsp-Y-Adda-Adda-d-Glu-N-methyldehydro-Ala), where X and Y represent variable l-amino acid residues. The amino acid Adda (3-amino-9-methoxy-2, 6, 8-trimethyl-10-phenyldeca-4, 6-dienoic acid) is considered to be responsible for the MC hepatotoxicity (Dawson, 1998). MC-LR (L and R designating leucine and arginine, respectively) was the first MC chemically identified, and it is the most toxic and most frequently found.

After ingestion, MC can penetrate into hepatocytes. Within the hepatocytes, MCs irreversibly inhibit protein phosphatases type 2A (PP2A) and 1 (PP1) (Dawson, 1998). External signs of poisoning, which include weakness, pallor, heavy breathing, vomiting and diarrhea, are then observed. MCs are potent tumor promoters, causing disruption of liver structure and function, haemorrhaging into the liver and death by respiratory arrest (Codd, 2000). Several cases of animal and human intoxication due to MCs have been reported. For example, in 1996, patients at a Brazilian hemodialysis center using municipal water contaminated with cyanotoxins were exposed to lethal levels of MCs. One hundred of the 131 patients developed acute liver failure and 52 of these victims are died due to hepatotoxin poisoning (Pouria et al., 1998; Jochimsen et al., 1998). In 2009, water pollution in Yancheng China affected the water supply system, which was closed for three days.

The toxicity and ubiquity of MCs necessitate the development of fast, sensitive and reliable methods to detect them. To guarantee water quality and to minimize the potential risk to human health, the World Health Organization (WHO) has recommended a maximum level of 1 $\mu g\ L^{-1}$ of MC-LR in drinking water (WHO, 1998). Accordingly, detection systems must be sensitive to MC concentrations below the limit established by the WHO. The simplest screening method is the mouse bioassay, which suffers from low sensitivity, specificity and ethical problems due to animal experimentation. In vitro cytotoxicity assays, based on morphological changes in cells after exposure to toxins, have been developed to provide a substitute for the mouse bioassay (Boaru, Dragos & Schirmer, 2006; Chong et al., 2000). These assays are easy to perform and economical but they are also subjective, time-consuming and confusing results may appear in the presence of toxin mixtures (Campas et al., 2007). Other cytotoxicity tests are based on the simple and sensitive analysis of the toxin effect on cells by measuring changes in O2 consumption by optical oxygen sensing technique (Jasionek et al., 2010). MCs are routinely analyzed using high-performance liquid chromatography (HPLC) coupled to mass spectrometry. These techniques allow highly selective identification and sensitive quantification of the different toxins present in a sample. However, they require expensive equipment, complex procedures, lengthy analysis times and trained personnel (Sangolkar, Maske & Chakrabarti, 2006; McElhiney & Lawton, 2005). An alternative and interesting approach is the use of biosensor for rapid, easy and sensitive detection of the toxin. A biosensor is a device composed of two intimately associated elements (Figure 1):

- A bioreceptor, that is an immobilized sensitive biological element (e.g. enzyme, DNA probe, antibody) recognizing the analyte (e.g. enzyme substrate, complementary DNA, antigen).

- A transducer, that is used to convert the (bio)chemical signal resulting from the interaction of the analyte with the bioreceptor into an electronic one. The intensity of generated signal is directly or inversely proportional to the analyte concentration. Biosensors can be based on electrochemical, gravimetric, calorimetric or optical detection. Electrochemical transducers (Thévenot et al., 1999) are classically used to develop biosensors (Ronkanein, Halsall & Heineman, 2010). These systems offer some advantages such as low cost, simple design or small dimensions.

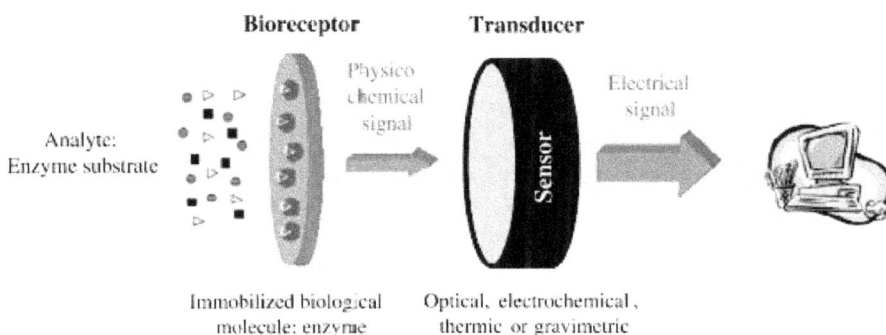

Figure 1. Scheme of a biosensor (Sassolas, Blum & Leca-Bouvier, 2012)

This review presents a state-of-the-art in electrochemical biosensors for the detection of microcystins. To clearly report the last advances, biosensors have been classified according to the immobilized recognition element. New trends in the field of microcystin analysis are also reviewed. Aptamers are shown as good candidates to replace the conventional antibodies and, thus, to be the biorecognition elements in more robust and stable biosensors for the detection of microcystin. Recent reports on the properties of nanomaterials show nanoparticles and nanotubes as promising tools to improve the efficiency of biosensors for the detection of microcystins.

2. Enzyme-inhibition based biosensors

2.1. Protein phosphatases

Enzyme inhibition-based methods have been widely used for the detection of MC. The main enzymatic method is based on the inhibition of protein phosphatases by the toxin. The enzyme inhibition can be detected by several methods such as colorimetry (Sassolas et al., 2011), fluorescence (Fontal et al., 1999) or electrochemistry (Campas et al., 2007; Campas et al., 2005; Szydlowska et al., 2006; Campas, Olteanu & Marty, 2008).

The commercially availability of protein phosphatases which avoids laborious purification makes the approach very attractive. Several assays and biosensors have been developed using PP2As purchased from Millipore (New York. USA). This enzyme is isolated as the heterodimer of 60 kDa (A) and 36 kDa (C) subunits from red blood cells. A French company called GTP Technology produced (by genetic engineering) PP2A that consists of a 39 kDa (tag included) human catalytic (C) subunit of the α-isophorm isolated from SF9 insect cells infected by baculovirus. Recently, PP2A from ZEU Immunotec (isolated from red blood cells) and a recombinant PP1 from CRITT (Toulouse, France) were also used for the development of colorimetric PP inhibition assay (Sassolas et al., 2011). Recently, the inhibition of three protein phosphatases (PP2As from ZEU Immunotec and GTP Technology, and PP1 from CRITT) by three MCs (LR, YR, RR) was investigated. It was demonstrated that the inhibition type in all cases was non-competitive (Covaci, Sassolas, Alonso, Munoz, Radu & Marty, 2012). The sensitivity of enzymes to MC can be drastically different. Thus, the choice of the enzyme is crucial to the performance of the system. A comparative study demonstrated that PP2A from ZEU Immunotec is more sensitive to MC than the other enzymes (Sassolas et al., 2011).

This enzymatic approach is limited due to the poor enzyme stability. To overcome this problem, enzymes can be immobilized. The choice of the immobilization technique is crucial for the performance of assays and biosensors.

2.2. Electrochemical enzyme-based biosensors for the detection of MCs

Marty's group developed electrochemical protein phosphatase-based biosensors for the detection of microcystin-LR (Campas et al., 2007; Szydlowska et al., 2006). The enzyme from Millipore was immobilized within a photopolymer formed on a screen-printed working electrode. The electrochemical measurement of the enzymatic activity was achieved using appropriate substrates electrochemically active after dephosphorylation by the enzyme. The enzymatic activity of PP2A is inhibited by the presence of MC, and hence the current intensity produced by the oxidation of electroactive product decreases proportionally to the toxin concentration. Several substrates were tested: catechol monophosphate, α-naphthyl phosphate and p-aminophenyl phosphate.

Real samples of cyanobacterial blooms from the Tarn River (Midi Pyrénées, France) have been analyzed using the developed amperometric biosensor (Campas et al., 2007). Electrochemical results were compared to those obtained by a conventional colorimetric protein phosphatase inhibition assay and HPLC. Despite the restricted sensitivity of the biosensor, potentially due to the electrode fouling by some cell extracts components, the applicability of the electrochemical system to rapidly assess the environmental and health risk due to MCs was demonstrated.

A signal amplification strategy based on enzymatic recycling was used to improve the sensitivity of the previously described biosensor (Campas et al., 2008). In this work, PP2A from Millipore was immobilized within a photopolymer. The detection principle was based on the dephosphorylation of non-electroactive p-aminophenyl phosphate (p-APP) by PP2A and the ability of diaphorase and NADH oxidase to recycle p-aminophenol in order to amplify the electrochemical signal arising from its oxidation (Figure 2). The amplification system allowed to improve the sensitivity of the biosensor. This strategy decreased the detection limit from 37.75 μg.L^{-1} to 0.05 μg.L^{-1} and enlarged the linear range by more than four orders of magnitude. The application of the amplification system to MC detection with a PP2A inhibition-based biosensor has resulted in a 755-fold lower detection limit, making the biosensor useful as a reliable screening tool to assess the water quality.

p-aminophenyl phosphate

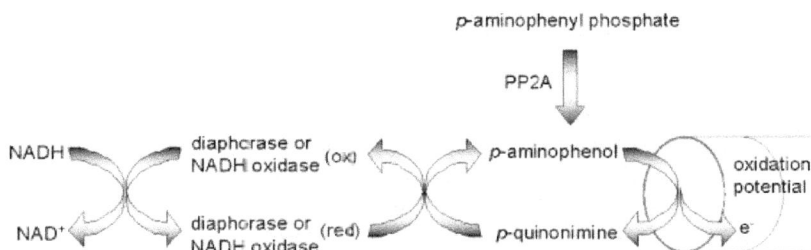

Figure 2. Scheme of the enzymatic signal amplification for the detection of the PP2A activity (Campas et al., 2008). The detection principle is based on the dephosphorylation of non-electroactive p-APP by PP2A, the oxidation of the corresponding electroactive p-aminophenol to p-iminoquinone (p-IQ) on the electrode surface and the regeneration of p-IQ by diaphorase, which requires NADH as substrate

3. Immunosensors

3.1. Principle

Immunosensors are characterized by the highly selective affinity interactions between antibodies (Ab) or antigens (Ag) immobilized on the transducer surface and their specific analytes, Ag or Ab respectively. Electrochemical immunosensors are obtained through the immobilization of the recognition element (Ag or Ab) on the electrode surface (Ricci, Adometto & Palleschi, 2012).

Different formats can be used to develop an electrochemical immunosensor. Due its small size (MW 900-1100 g.mol^{-1}), sandwich assays are not possible for MC detection. Two approaches could be considered when dealing with competitive immunosensors. A first one in which immobilized antibodies react with free antigens in competition with labeled antigens (Campas & Marty, 2007). A second one, using immobilized antigens and labeled antibodies, is generally preferred and prevents all the problems related to antibody immobilization (loss of affinity, orientation of the immobilized protein) (Ricci et al., 2012). Electrochemical detection of immunoreactions can also be carried out directly without label. In this case, the immunosensor measures changes in electronics or interfacial properties due to the Ag/Ab complex formation on the electrode surface.

3.2. Label-based electrochemical immunosensors

Marty's group developed the first immunosensor for the MC analysis (Campas & Marty, 2007) This amperometric immunosensor was based on the affinity between the cyanotoxin and the corresponding monoclonal or polyclonal Abs. In this work, Abs were immobilized on a SPE. The toxin present in a sample and MC-HRP conjugate competed for binding to the immobilized anti-MC Ab. The electrochemical detection required the use of 5-methyl-phenazinium methyl sulfate (MPMS) as a redox mediator in order to provide electrical contact between enzymatic label and the electrode surface. The more MC present in the sample, the less the electrochemical signal measured. The monoclonal Ab system provides lower detection limit. In this case, the MC-LR could be detected between 0.1 ng/L^{-1} and 100 µg/L^{-1}. Although the monoclonal Ab sensor was sensitive, this system had a low reproducibility and, consequently was not enough reliable. Despite the potential matrix electrochemical effects, the analysis of algal samples with both immunosensors and the comparison of the results with those obtained with a colorimetric enzymatic assay and HPLC validate the applicability of the developed devices as screening tools for fast and reliable cyanotoxin detection.

In the last few years, carbon nanomaterials have been used in sensitive detection of various analytes. Recently, single-walled carbon nanohorns (SWNHs) were used for the development of an electrochemical immunosensor for rapid detection of MC-LR (Zhang et al., 2010). SWNHs are spherical aggregates of thousands of graphitic tubule closed ends with cone shaped horns. The functionalization of SWNHs was performed by covalently binding MC-LR to the carboxylic group on the cone-shaped tips of SWNHs. Competition of HRP-labeled MC-LR Ab for free and immobilized toxin was subsequently performed. This immunosensor exhibited a wide linear response to MC-LR ranging from 0.05 $\mu g.L^{-1}$ to 20 $\mu g.L^{-1}$ with a detection limit of 0.03 $\mu g.L^{-1}$. This method showed good accuracy and reproducibility. This immunosensor was used for the analysis of polluted water samples and the results are in good agreement with those obtained with HPLC.

However, these methods require an enzyme label. The analyses are rather complicated with relatively high cost.

3.3. Label-free immunosensors

The development of label-free immunosensors represents an attractive approach for detecting affinity interactions. These systems exploit the unique properties of nanomaterials. To further enhance capability of a biosensor, immobilization using nanomaterials is of considerable interest. The field of environmental diagnostics has been interested in using NPs for analyzing toxins (Wang et al., 2010). Metal nanoparticles are generally defined as isolable particles between 1 and 50 nm in size, that are prevented from agglomerating by protecting shells. Owing to their small size, nanoparticles have physical, electronic and chemical properties that are different from those of bulk metals. Such properties strongly depend on the number and kind of atoms that make up the particle (Wang, 2005a). Nanoparticles have been exploited as biomolecule immobilization supports because of their large surface-to-volume ratio, high surface reaction activity for biomolecule loading and high catalytic activity. On the one hand, a high number of biomolecules can be immobilized on NPs, retaining their biological activity. On the other hand, electron transfer between biomolecules and electrode surfaces is promoted (Campas, Garibo & Prieto-Simon, 2012). The properties of NPs have been exploited for the development of sensitive immunosensors. For instance, a label-free capacitive immunosensor for the detection of MC-LR was developed (Loyprasert et al., 2008). Anti-MC-LR Ab was immobilized on silver NPs bound to a self-assembled thiourea monolayer formed on the working electrode. NPs were incorporated into modified electrodes to enhance response and achieve a more sensitive system. Capacitive immunosensors measure the changes in dielectric properties when an Ag/Ab complex is formed on the surface of an electrode. Under optimum conditions, the detection limit was 7 $pg.L^{-1}$. The immobilized anti-MC-LR Ab on self-assembled thiourea monolayer incorporated with silver nanoparticles was stable and good reproducibility of the signal could be obtained up to 43 times with a R.S.D. of 2.1 %. The immunosensor was applied to analyze MC-LR in water samples and the results were in good agreement with those obtained by HPLC. A label-free impedimetric immunosensor for the detection of MC-LR was also developed by immobilizing Ab on gold Nps/L-cysteine coated electrode (Sun, et al., 2010). Under optimal conditions, MC-LR could be determined with a detection limit of 18.2 $ng.L^{-1}$. Moreover, the immunosensor exhibited a long-term stability and good reproducibility of the signal could be obtained up to 42 times with a R.S.D. of 3.58 %. The same strategy was also used to develop a label-free amperometric immunosensor for the detection of MC-LR in water (Tong et al., 2011). In this case, the detection limit was 20 ng L^{-1}.

The unique chemical and physical properties of carbon nanotubes (CNTs) have paved the way to new electrochemical biosensors (Wang, 2005b; Viswanathan & Radecki, 2008; Balasubramanian & Burghard, 2006). CNTs can be described as a graphite sheet rolled up into a nanoscale-tube (Single-wall carbon nanotubes, SWCNT) or with additional graphene tubes around the core of a SWCNT (multi-wall CNTs, MWCNTs) (Trojanowicz, 2006). A SWCNT-coated paper was developed for the detection of MC-LR (Wang et al., 2009). First, Ab was dispersed together with SWCNTs. Then, the dispersion was used to dip-coat the paper rendering it conductive. The obtained SWCNT-coated paper was used as working electrode. The interaction MC-LR/Ab induced a change in conductivity of SWCNT-coated paper, which was used to detect the toxin in the water. The detection limit was found to be 0.6 $\mu g.L^{-1}$.

4. DNA sensors

The use of DNA recognition layers has been extensively explored in the field of analytical chemistry due to their wide range of physical, chemical and biological activities (McGown et al., 1995). An oligonucleotide with a known sequence of bases or a fragment of DNA is used as sensing element in DNA biosensors (Sassolas, Leca-Bouvier, & Blum, 2008). The DNA biosensors (also called genosensors) are either based on the hybridization of complementary strands of DNA or could be used as highy specific receptor for many target molecules (Singh et al., 2012). Given the high affinity and specificity for the target molecules, DNA biosensors have potential applications in a variety of detection and diagnostic systems and can thus be considered as a valid alternative to antibodies or other bio-receptors, for the development of biosensors. DNA offers many advantages compared to antibody in the design of biosensors. In contrast to antibody, conformational changes of DNA can also be explored to monitor molecular recognition event. The handling of biosensors based on DNA is much easy than that of immunosensors, since proteic nature of antibody limits their application to physiological conditions. The stability of DNA biosensor is higher than that of antibody based devices. DNA biosensors can stand to high temperature, low pH and high level of organic solvent that denatures antibodies. Furthermore, immobilized DNA can be regenerated easily by changing the temperature or pH or by the addition of chaotropic agent n case of DNA biosensors. Other advantage is the easy conjugation of the labelled molecules to DNA and also easy immobilization of DNA on different transducer surfaces in the design of biosensors. Furthermore, the small size and versatility of DNA strands make them suitable to immobilize in high-density monolayer onto the electrode surface, which is of vital importance to miniaturise the biosensors (Wang et al., 1997; Bagni, et al., 2006).

The detection techniques play very important role in the design of biosensors, and should be selected according to the requirement of particular application. Among the various sensing device designed so far, electrochemical DNA biosensors have attracted more attention due to their high sensitivity and rapid response. In addition electrochemical techniques are suitable for miniaturization and have the potential to simplify the nucleic acid analysis using low cost electronics.

Yan *el al.* (2001) developed an electrochemical biosensor for voltammetric detection of gene sequence related to genera of cynobacteria, *Microcystis* spp. The sensor was based on the immobilization of a complementary 17-mer DNA probe on a gold electrode through specific adsorption. The target gene in the solution was determined by the use of methylene blue and ruthenium bipyridine as electrochemical indicators. $Ru(bpy)_3^{2+}$ interacts with guanines of ssDNA whereas the formation of the double helix precludes the collision of $Ru(bpy)_3^{2+}$ with guanine bases. When $Ru(bpy)_3^{2+}$ oxidizes guanine, a high catalytic cuurent is measured after hybridization. Methylene blue is used as an electrochemical intercalator to monitor the DNA hybridization because ssDNA and dsDNA have different affinities for it (Sassolas, Leca-Bouvier et al., 2008). The anodic peak currents of $Ru(bpy)_3^{2+}$ were linearly related to the concentration of the target oligonucleotide sequence in the range The detection limit of this approach was 9.0×10^{-11} M (Yan et al., 2001). Similarly, Erdem et al. (2002) described an electrochemical biosensor for the voltammetric detection of DNA sequences related to *Microcystis* spp. A specific DNA probe was designed and immobilized on the carbon paste electrode. Differential pulse voltammetry (DPV) and cyclic voltammetry (CV) in the presence of methylene blue and tris ruthenium were used to evaluate the hybridization process between the target sequence and the immobilized DNA probe. The system was used to detect Microcystis spp from real tap water and river water (Erdem et al., 2002). Recently, Lan et al. (2010) developed a disposable electrochemical DNA biosensor for in situ determination of *Microcystis* spp. The DNA probe, complementary to target DNA, was immobilized on the screen printed carbon electrode (SPCE) surface by the use of gold nanoparticles (AuNPs). The AuNPs were used to enhance the immobilization process and subsequently to increase the system sensitivity. Methylene blue was used as redox indicator and different immobilization steps were characterized by CV and DPV (Lan et al., 2010). Although electrochemical DNA biosensors based on the use of redox indicator provided reliable and precise information in mutation detections, redox probes require high potential that are prone to interferences and often destroy the hybrid double strand structure. To overcome this problem, an electrochemical DNA biosensor based on the concept of metal-enhanced detection for the determination of *Microcystis* spp was developed (Owino, Mwilu & Sadik, 2007). The biosensor was constructed by immobilizing 17-mer DNA probe on a gold electrode via avidin-biotin chemistry. Electrochemical reduction and oxidation of DNA-captured Ag^+ ions provided the detection signal with a detection limit of 7×10^{-9} M.

5. Aptamer-based biosensors

Aptamers, a new class of molecules, have been appeared as promising recognition tools for analytical applications. Aptamers are short single stranded oligonucleotides, either DNA or RNA that fold into well-defined 3D structures and bind to their ligand by complementary shape interactions, with antibody-like binding ability. They are engineered through an *in vitro* selection procedure, also called SELEX (Systematic Evolution of Ligands by EXponential enrichment), which was first reported in 1990 (Ellington & Szostak, 1990; Tuerk & Gold, 1990). The SELEX process is a technique for screening very large combinatorial libraries of oligonucleotides by an iterative process of *in vitro* selection and amplification. In the screening process, a random sequence oligonucleotide library is incubated with the target of interest. Sequences that bind to the target are separated from the unbound species using a suitable partitioning method, and then, the sequence of these candidates is amplified using polymerase chain reaction (PCR). This population of selected sequences represents a mixture of oligomers with variable affinity towards the target analyte. The single-strand population obtained after the purification step is incubated with a fresh sample of the target for the next round of selection. Iteration of the above protocol results in the isolation of a pool of nucleotide sequences displaying sequential motifs, which after 8-15 iterative SELEX runs converge to one or a few binding sequences. Once the sequence is identified, an aptamer is produced by chemical synthesis (Stoltenburg, Reinemann & Strehlitz, 2007).

Aptamers hold significant advantages over other bioreceptor molecules. As they are chemically synthesized, their production does not require the use of animal and is therefore less expensive and tedious. Aptamers can be a so easily labeled with a wide range of reporter molecules such as fluorescent dyes, enzymes, biotin, or aminated compounds, enabling the design of a variety of detection methods. Due to its many advantages, numerous aptamer-based biosensors have been developed for the detection of a wide range of targets (Sassolas, Blum & Leca-Bouvier, 2011; Sassolas, Blum & Leca-Bouvier, 2008; Hianik & Wang, 2009).

The selection of an appropriated aptamer, and its use in the development of aptasensors could offer various advantages over the existing methods for microcystin detection. A surface Plasmon resonance (SPR) aptasensor based on anti-microcystin DNA aptamer was developed by Nakamura et al. in 2001. The sensitivity of the aptasensor was not as high compared with the methods reported previously, suggesting further improvement in the SELEX process for microcystin molecules (Nakamura et al., 2001). Recently, a RNA aptamer specific for microcystin-LR has also been selected (Gu & Famulok, 2004; Hu et al., 2012). Currently, the potential of aptamers for the microcystin detection has not still been exploited but electrochemical aptamer-based biosensors could be an alternative to the conventional methods of toxin analysis.

6. Conclusion

Biosensors are good candidates for the environmental monitoring. They exploit the remarkable specificity of recognition elements to design efficient analytical tools that can detect the presence of microcystins in water. Several configurations that depend on the type of recognition elements and the immobilization technique can be envisaged for the development of biosensors. The biological elements used for toxin analysis are classically enzymes and antibodies. In the last decade, aptamers have been used as new molecular recognition elements to develop biosensors. However, unique properties of aptamers have not been yet exploited for the microcystin analysis. We believe that these recognition elements could be used, in the near future, for the development of efficient electrochemical aptasensors allowing the detection of microcystins.

Nanotechnology is playing an important role in the development of efficient biosensors for the toxin detection. Different types of nanomaterials (e.g. nanoparticles and nanotubes) with different properties have been used. They offer exciting new opportunities to improve the performance of electrochemical biosensors for the detection of microcystins.

The use of biosensors in environmental field is still limited in comparison to medical applications. Most commercial biosensors are for medical applications, whereas only few are adapted for the environmental monitoring. Even if some commercial kits are available for the analysis of DSP toxins, there is still a challenge to develop improved and more reliable devices allowing the analysis of microcystins in water.

References

Bagni, G., Osella D., Sturchio E., & Mascini M. (2006). Deoxyribonucleic acid (DNA) biosensors for environmental risk assessment and drug studies. *Analytica Chimica Acta, 573-574,* 81-89. http://dx.doi.org/10.1016/j.aca.2006.03.085

Balasubramanian, K., & Burghard, M. (2006). Biosensors based on carbon nanotubes. *Analytical and Bioanalytical Chemistry, 385,* 452-468. http://dx.doi.org/10.1007/s00216-006-0314-8

Boaru, D.A., Dragos, N., & Schirmer, K. (2006). Microcystin-LR induced cellular effects in mamalian and fish primary hepatocyte cultures and cell lines: a comparative study. *Toxicology, 218,* 134-148. http://dx.doi.org/10.1016/j.tox.2005.10.005

Campas, M., Szydlowska, D., Trojanowicz, M. & Marty, J.L. (2005). Towards the protein phosphatase-based biosensor for microcystin detection. *Biosensors and Bioelectronics, 20,* 1520-1530. http://dx.doi.org/10.1016/j.bios.2004.06.002

Campas, M., Szydlowska, D., Trojanowicz, M., & Marty, J.L. (2007). Enzyme inhibition-based biosensor for the electrochemical detection of microcystins in natural blooms of cyanobacteria. *Talanta, 72,* 179-186. http://dx.doi.org/10.1016/j.talanta.2006.10.012

Campas, M., & Marty, J.-L. (2007). HIghly sensitive amperometric immunosensors for microcystin detec tion in algae. *Biosensors and Bioelectronics, 22,* 1034-1040. http://dx.doi.org/10.1016/j.bios.2006.04.025

Campas, M., Garibo, D., & Prieto-Simon, B. (2012). Novel nanobiotechnological concepts in electrochemical biosensors for th analysis of toxins. *Analyst, 137,* 1055-1067. http://dx.doi.org/10.1039/c2an15736e

Campas, M., Olteanu, M.G., & Marty, J.-L. (2008). Enzymatic recycling for signal amplification: improving microcystin detection with biosensors. *Sensors and Actuators B, 129,* 263-267. http://dx.doi.org/10.1016/j.snb.2007.08.009

Chong, M.W.K., Gu, K.D., Lam, P.K.S., Yang, M., & Fong, W.F. (2000). Study on the cytotoxicity of microcystin-LR on cultured cells. *Chemosphere, 41,* 143-147. http://dx.doi.org/10.1016/S0045-6535(99)00402-6

Codd, G.A. (2000). Cyanobacterial toxins, the perception of water quality, and the prioritisation of eutrophication control. *Ecological Engineering, 16,* 51-60. http://dx.doi.org/10.1016/S0925-8574(00)00089-6

Covaci, O.I., Sassolas, A., Alonso, G.A., Munoz, R., Radu, G.L., & Marty, J.L. (2012). Highly sensitive detection and discrimination of LR and YR microcystins based on protein phosphatases and an artificial neural network. *Analytical and Bioanalytical Chemistry, 404,* 711-720. http://dx.doi.org/10.1007/s00216-012-6092-6

Dawson, R.M. (1998). The toxicology of microcystins. Toxicon, 36(7), 953-962. http://dx.doi.org/10.1016/S0041-0101(97)00102-5

de Figueiredo, D.R., Azeiteiro, U.M., Esteves, S.M., Gonçalves, F.J.M., & Pereira, M.J. (2004). Microcystin-producing blooms-a serious global public health issue. *Ecotoxicology and Environmental Safety, 59,* 151-163. http://dx.doi.org/10.1016/j.ecoenv.2004.04.006

Ellington, A.D., & Szostak, J.W. (1990). In vitro selection of RNA molecules that bind specific ligands. *Nature, 346,* 818-822. http://dx.doi.org/10.1038/346818a0

Erdem, A., Kerman, K., Meric, B., Ozkan, D., Kara, P., & Ozsoz, M. (2002). DNA biosensor for microcystis spp. sequence detection by using methylene blue and ruthenium complex as electrochemical hybridization labelsa. *Turkish Journal of Chemistry, 26,* 851-862.

Fontal, O.I., Vieytes, M.R., Baptista de Sousa, J.M.V., Louzao, M.C., & Botana, L.M. (1999). A fluorescent micropate assay for microcystin-LR. *Analytical Biochemistry, 269,* 289-296. http://dx.doi.org/10.1006/abio.1999.3099

Gu, K.D., & Famulok, M. (2004). In vitro selection of specific aptamers against microcystin-LR. *Chinese Journal Prev. Med., 38,* 369-373

Hianik, T., & Wang, J. (2009). Electrochemical aptasensors-recent achievements and perspectives. *Electroanalysis,* 1223-1235. http://dx.doi.org/10.1002/elan.200904566

Hu, X., Mu, L., Wen,. J., & Zhou, Q. (2012). Immobilized smart RNA on graphene oxide nanosheets to specifically recognize and adsorb trace peptide toxins in drinking water. *Journal of Hazardous Materials, 213-214,* 387-392. http://dx.doi.org/10.1016/j.jhazmat.2012.02.012

Jasionek, G., Zhdanov, A., Davenport, J., Blaha, L., & Papkovsky, D.B. (2010). Mitochondrial toxicity of microcystin-LR on cultured cells: application to the analysis of contaminated water samples. *Environ. Sci. Technol.. 44,* 2533-2541. http://dx.doi.org/10.1021/es903157h

Jochimsen, E.M., Carmichael, W.W., An, J., Cardo, D., Cookson, S.T., & Holmes, C.E.M. (1998). Liver failure and death after exposure to microcystins at a hemodialysis centre in Brazil. *The New England Journal of Medecine, 338(13),* 873-878. http://dx.doi.org/10.1056/NEJM199803263381304

Lan, M., Chen, C., Zhou, Q., Teng, Y., Zhao, H. & Niu, H. (2010). Voltammetric detection of microcystis genus specific-sequnce with disposable screen-printed electrode modified with gold nanoparticles. *Advanced Materials Letters, 1(3),* 217-224. http://dx.doi.org/10.5185/amlett.2010.7144

Loyprasert, S., Thavarungku , P., Asawatreratanakul, P., Wongkittisuksa, B., Limsakul, C., & Kanatharana, P. (2008). Label-free capacitive immunosensor for microcystin-LR using self-assembled thiourea monolayer incorporated with Ag nanoparticules on gold electrode. *Biosensors and Bioelectronics, 24,* 78-86. http://dx.doi.org/10.1016/j.bios.2008.03.016

McElhiney, J., & Lawton, L.A. (2005). Detection of the cyanobacterial hepatotoxins microcystins. *Toxicology and Applied Pharmacology, 203,* 219-230. http://dx.doi.org/10.1016/j.taap.2004.06.002

McGown, L.B., Joseph, M.J., Pitner, J.E., Vonk ,G.P., & Linn, C.P. (1995). The nucleic acid ligand. A new tool for molecular recognition. *Analytical Chemistry, 67(21),* 663A-668A.

Nakamura, C., Kobayashi, T., Miyake, M., Shirai, M., & Miyake, J. (2001). Usage of a DNA aptamer as a ligand targeting microcystin. Molecular Crystals and Liquid Crystals Science and Technology. *Section A, Molecular Crystals and Liquid Crystals, 371,* 369-374. http://dx.doi.org/10.1080/10587250108024762

Owino, I.O., Mwilu, S.K., & Sadik, O.A. (2007). Metal-enhanced biosensor for genetic mismatch detection. *Analytical Biochemistry, 359,* 8-17. http://dx.doi.org/10.1016/j.ab.2007.06.046

Pouria, S., de Andrade, A., Barbosa, I., Cavalcanti, R.L., Barreto, V.T.S., Ward, C.J., et al. (1998). Fatal microcystin intoxication on haemodialysis unit in Caruaru, Brazil. *The Lancet, 352,* 21-26. http://dx.doi.org/10.1016/S0140-6736(97)12285-1

Ricci, F., Adometto, G., & Palleschi, G. (2012). A review of experimental aspects of electrochemical immunosensors. *Electrochimica Acta.* http://dx.doi.org/10.1016/j.electacta.2012.06.033

Ronkanein, N.J., Halsall, H.B., & Heineman, W.R. (2010). Electrochemical biosensors. *Chemical Society Reviews, 39,* 1747-1763. http://dx.doi.org/10.1039/b714449k

Sangolkar, L.N., Maske, S.S., & Chakrabarti, T. (2006). Methods for determining microcystins (peptide hepatotoxins) and microcystin-producing cyanobacteria. *Water Research, 40,* 3485-3496. http://dx.doi.org/10.1016/j.watres.2006.08.010

Sassolas, A., Blum, L.J., & Leca-Bouvier, B.D. (2008). Electrochemical aptasensors. Electroanalysis, 21(11), 1237-1250. http://dx.doi.org/10.1002/elan.200804554

Sassolas, A., Blum, L.J., & Leca-Bouvier, B.D. (2011). Optical detection systems using immobilized aptamers. *Biosensors and Bioelectronics, 26(9),* 3725-3736. http://dx.doi.org/10.1016/j.bios.2011.02.031

Sassolas, A., Blum, L.J., & Leca-Bouvier, B.D. (2012). Immobilization strategies to develop enzymatic biosensors. *Biotechnology Advances, 30(3),* 489-511. http://dx.doi.org/10.1016/j.biotechadv.2011.09.003

Sassolas, A., Catanante, G., Fournier, D., & Marty, J.L. (2011). Development of a colorimetric inhibition assay for microcystin-LR detection: comparison of the sensitivity of different protein phosphatases. *Talanta, 85,* 2498-2503. http://dx.doi.org/10.1016/j.talanta.2011.07.101

Sassolas, A., Leca-Bouvier, B.D., & Blum, J.J. (2008). DNA biosensors and microarrays. *Chemical Reviews, 108(1),* 109-139. http://dx.doi.org/10.1021/cr0684467

Singh, S., et al. (2012). Recent trends in development of biosensors for detection of microcystin. *Toxicon, 60(5),* 878-894. http://dx.doi.org/10.1016/j.toxicon.2012.06.005

Stoltenburg, R., Reinemann, C., & Strehlitz, B. (2007). SELEX-a (r)evolutionary method to generate high-affinity nucleic acid ligands. *Biomolecular Engineering, 24,* 381-403. http://dx.doi.org/10.1016/j.bioeng.2007.06.001

Sun, X., Shi ,H., Wang, H., Xiao, L., Li,, L., et al. (2010). A simple, highly sensitive, and label-free impedimetric immunosensor for detection of microcystin-LR in water. *Analytical Letters, 43,* 533-544. http://dx.doi.org/10.1080/00032710903406912

Szydlowska, D., Campas, M., Marty, J.L., & Trojanowicz, M. (2006). Catechol monophosphate as a new substrate for screen-printed amperometric biosensors with immobilized phsophatases. *Sensors and Actuators, 113,* 787-796. http://dx.doi.org/10.1016/j.snb.2005.07.041

Thévenot, D.R., Toth, K., Durst, R.A., & Wilson, G.S. (1999). Electrochemical biosensors: recommended definctions and classification. *Pure Appl. Chem., 71(12),* 2333-2348. http://dx.doi.org/10.1351/pac199971122333

Tong, P., Tang, S., He, Y., Shao, Y., Zhang, L., & Chen, G. (2011). Label-free immunosensing of microcystin-LR using a gold electrode modified with gold nanoparticles. *Microchimica Acta, 173,* 299-305. http://dx.doi.org/10.1007/s00604-011-0557-8

Trojanowicz, M. (2006). Analytical applications of carbon nanoubes: a review. *Trends in Analytical Chemistry, 25(5),* 480-489. http://dx.doi.org/10.1016/j.trac.2005.11.008

Tuerk, C., & Gold, L. (1990). Systematic evolution of ligands by exponential enrichment: RNA ligands to bacteriophage T4 DNA polymerase. *Science, 249,* 505-510. http://dx.doi.org/10.1126/science.2200121

Viswarathan, S., & Radecki, J. (2008). Nanomaterials in electrochemical biosensors for food analysis-a review. *Polish Journal of Food and Nutrition Sciences, 58(2),* 157-164.

Wang, J., Rivas, G., Cai, X., Palecek, E., Nielsen, P., & Shiraishi, H. (1997). DNA electrochemical biosensors for environmental monitoring. A review. *Analytica Chimica Acta, 347(1-2),* 1-8. http://dx.doi.org/10.1016/S0003-2670(96)00598-3

Wang, J. (2005a). Nanomaterial-based electrochemical biosensors. *Analyst, 130,* 421-426. http://dx.doi.org/10.1039/b414248a

Wang, J. (2005b). Carbon-nanotube based electrochemical biosensors: a review. *Electroanalysis, 17(1),* 7-14. http://dx.doi.org/10.1002/elan.200403113

Wang, L., Chen. W., Xu. D., Shim. B.S., Zhu. Y., Sun, F., et al. (2009). Simple, rapid, sensitive, and versatile SWNT-paper sensor for environmental toxin detection competitive with ELISA. *Nano Letters, 9(12),* 4147-4152. http://dx.doi.org/10.1021/nl902368r

Wang, L., Ma, W., Xu, L., Chen, W., Zhu, Y., Xu, C., Kotov, N.A., et al (2010). Nanoparticle-based environmental sensors. *Materials Science and Engineering R, 70,* 265-274. http://dx.doi.org/10.1016/j.mser.2010.06.012

WHO (1998). Guidelines for drinking water quality. Health Criteria and Other supporting information. Second edition. Volume 2, Addendum. Geneva, 95-110.

Yan, F., Erdem, A., Meric, B., Kerman, K., Ozsoz, M., & Sadik, O.A. (2001). Electrochemical DNA biosensor for the detection of specific gene related to Microcystis species. *Electrochemistry Communications, 3(5),* 224-228. http://dx.doi.org/10.1016/S1388-2481(01)00149-7

Zhang, J., Lei, J., Xu, C., Ding, L., & Ju, H. (2010). Carbon nanohorn sensitized electrochemical immunosensor for rapid detection of microcystin-LR. *Analytical Chemistry. 82,* 1117-1122. http://dx.doi.org/10.1021/ac902914r

Chapter 5

Phenolic compounds determination in water by enzymatic-based electrochemical biosensors

Juan C. Gonzalez-Rivera[1], Jorge E. Preciado[2], Johann F. Osma[3]

[1]Microelectronics Research Center (CMUA), Department of Electrical and Electronics Engineering, Universidad de los Andes, Cra 1 E No. 19 A – 40, Bogota, Colombia.

[2]Universidad de los Andes, Cra 1 E No. 19 A – 40, Bogota, Colombia.

[3]Microelectronics Research Center (CMUA), Department of Electrical and Electronics Engineering, Universidad de los Andes, Cra 1 E No. 19 A – 40, Bogota, Colombia.

jc.gonzalez141@uniandes.edu.co, je.preciado40@uniandes.edu.co, jf.osma43@uniandes.edu.co

Doi: http://dx.doi.org/10.3926/oms.148

Referencing this chapter

Gonzalez-Rivera, J.C., Preciado, J.E., Osma, J.F. (2014). Phenolic Compounds Determination in Water by Enzymatic-based Electrochemical Biosensors. In M. Stoytcheva & J.F. Osma (Eds.). *Biosensors: Recent Advances and Mathematical Challenges*. Barcelona: España, OmniaScience. pp. 111-127.

1. Introduction

Nowadays, phenols are essential compounds broadly employed in the chemical, petrochemical, pharmaceutical, pesticide, pulp and paper, textile, metallurgic, resin and plastic, and pulp and paper industries. These chemicals are commonly used in the manufacture and processing of plastics and plasticizers, resins, explosives, drugs, detergents, paper, herbicides, insecticides, algaecides, bactericides, molluscicides, fungicides, preservatives, dyes, paints, lubricants, and fuel and solid waste combustion (Glezer, 2003; Michałowicz & Duda, 2007). In fact, the global capacity, only for phenol production, was about 9.9 million tons in 2008 (Weber & Weber, 2010). Nonetheless, most phenolic compounds are distinguished by their toxic, noxious, mutagenic and carcinogenic activity (Michałowicz & Duda, 2007). These recalcitrant pollutants accumulate over time and are found in food, potable water, sediments and soil.

Currently, some organizations have developed technical standards to determinate phenols mainly in water. The US Environmental Protection Agency (EPA), American Section of the International Association of Testing Materials (ASTM), and the International Organization for Standardization (ISO) have established several procedures for phenolic compounds determination in drinking, ground, surface, and saline waters, and domestic and industrial wastes by colorimetric, gas and liquid chromatography, capillary electrophoresis and their variations (Karim & Fakhruddin, 2012). Some of these procedures are EPA Method 420.4 (U.S. Environmental Protection Agency, 1993), EPA Method 528 (U.S. Environmental Protection Agency, 2000), EPA Method 604, (U.S. Environmental Protection Agency, 1996a, 2000) EPA Method 625 (U.S. Environmental Protection Agency, 1996b), ISO 8165-1:1992 (International Organization of Standardization, 1992), ISO 8165-2:1999 (International Organization of Standardization, 1999), and ASTM D1783-01 (ASTM International, 2012). Even though these standardized methods are able to obtain accurate results for a wide range of phenolic compounds, conventional approaches are time-consuming and cost-intensive. Furthermore, they require large volumes of toxic organic solvents such as methylene chloride, acetone, and methanol. Consequently, there is a demand for the development of reliable, portable, sensitive, simple and cost-effective methods for fast detection of phenolic compound.

Enzymatic-based biosensors have grown as a promising technology in the detection field. Enzymes are the most widely used biological sensing element in the fabrication of biosensors. The main advantages of biosensors are high catalytic activity, substrate selectivity, moderate operational potentials, high sensitivity, specificity, and applicability to multi component solutions in situ (Borgmann, Schulte, Neugebauer & Schuhmann, 2011; Karim & Fakhruddin, 2012).

This chapter seeks to provide an overview of recent advances on enzymatic-based biosensors for phenolic compound detection. In particular, specific redox-enzymes such as laccase, tyrosinase and peroxidase were described and compared. Additionally, a review of the main phenolic compounds and their toxicity for human health is highlighted to understand the importance of biosensor development.

2. Phenols

Phenolic compounds are based on organic molecules which contain a hydroxyl group bounded directly to an aromatic ring. This ring provides their characteristic stability due to electron resonance. As aromatic structure is energetically favorable, there is a preference for electrophilic aromatic substitution reactions such as chlorination, sulphonations, nitration, and nitrosation (Nguyen, Kryachko & Vanquickenborne, 2003).

The term phenols cover a very large group of chemical compounds, the most relevant and common are: (1) chlorophenols, (2) nitrophenols, (3) catechols, (4) chlorocatechols, (5) alkylphenols, (6) bisphenols, and (7) aminophenols. Figure 1 shows some of these compounds.

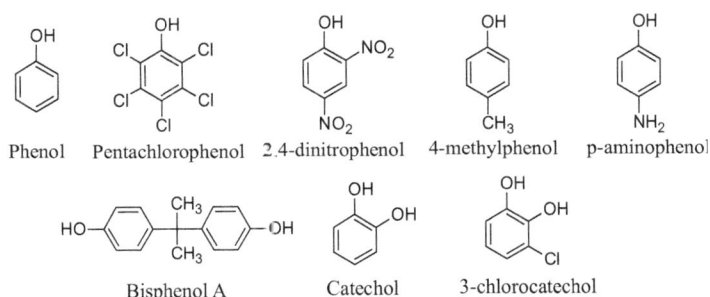

Figure 1. Examples some phenolic compounds

The US Environmental Protection Agency (EPA) as wells as The European Union (EU) have classified phenolic compounds as priority pollutants due to their strong toxic influence (Glezer, 2003). Not only chemical processing and manufacturing plant workers are exposed to phenolic compounds, but also the general population. Phenols are present in the environment due to the current industry activity, mostly in water, sewage, waste, soil, air, and even food. Moreover, their toxicity activity increases owing to their poor biodegradability (Glezer, 2003).

Phenol poisoning by skin absorption, inhalation of vapors or ingestion cause accumulation and damaging of brain, kidneys, liver, muscle, and eyes as well as necrosis. Phenol skin contact in high concentrate solutions (60%-90%) and a dose of 1 g of phenol may be lethal for an adult (Karim & Fakhruddin, 2012; Michałowicz & Duda, 2007; Services U.S. Department of Health and Human, 2008). Chlorophenols are subjected to fast skin and mucous membrane absorption. Burning pain, white necrotic lesions, vomiting, headache, irregular pulse, temperature decrease, muscle weakness, convulsions and death are typical characteristics of chlorophenols exposure The acceptable daily intake for pentachlorophenol was established for 16 μg for a man of 70 kg of bodyweight. Nitrophenol exposure undergoes to similar consequences as chlorophenols. In fact, compounds like 2,4-dinitrophenol were used as a slimming drug and food additive until it was removed from the market because of numerous cases of chronic heat, depression and deaths. The median lethal dose of this compound is considered to be 14 to 35 mg kg^{1} (Michałowicz & Duda, 2007). Industrial workers are highly exposed to methylphenols, especially 4-methylphenol, which causes burning pain, abdominal pain, headache, weak irregular pulse, hypotension, fall of body temperature, dark colored urine, shock, paralysis, coma and death. The median lethal dose for rats is 207 mg kg^{-1}. Bisphenol A might be a factor of decreasing seminal fluid, increase rates of breast cancer in women and a decrease of plasma luteinizing hormone level. Furthermore, aminophenol toxicity entails semiquinones and superoxide radical generation, which damage the cell's biomolecules. For p-aminophenol, the lethal dose for an adult man is estimated to be 50 to 500 mg kg^{-1} of bodyweight (Michałowicz & Duda, 2007).

Phenol, catechols, nitrophenols, aminophenols, bisphenols show mutagenic activity as they damage and inhibit DNA synthesis and replication inside cells. In addition, carcinogenic activity, influencing tumor growth processes, has been demonstrated for some phenolic compounds such as 2-chlorophenol, 2,4-dichlorophenol, 2,4,6-trichlorophenol, catechol, p-cresol, 2,4-dimethylphenol, 4-methoxyphenol, and bythylhydroxyanisole (Michałowicz & Duda, 2007).

3. Biosensors

The first reported biosensor was an oxygen enzyme-based electrode developed in 1962 by (Clark & Lyons, 1962). The term biosensor was introduced for the first time by (Cammann, 1977). According to the IUPAC, a biosensor is a device that transforms biochemical information into an analytical useful signal through a physicochemical transducer and a biological recognition system (receptor) that selectively reacts with the analyte of interest (Thévenot, Toth, Durst & Wilson, 2001).

A typical biosensor set-up, illustrated in Figure 2, consists on (1) a biorecongnition element (enzyme, antibody, nucleic acid, living cells or microorganism), (2) a transducer that enables to transfer the output signal from the receptor to a measurable response, mostly an electric signal, (3) an amplifier that magnifies the output of the transducer, and (4) a data acquisition element which converts the sensor signal into digital values (Grieshaber, MacKenzie, Vörös & Reimhult, 2008).

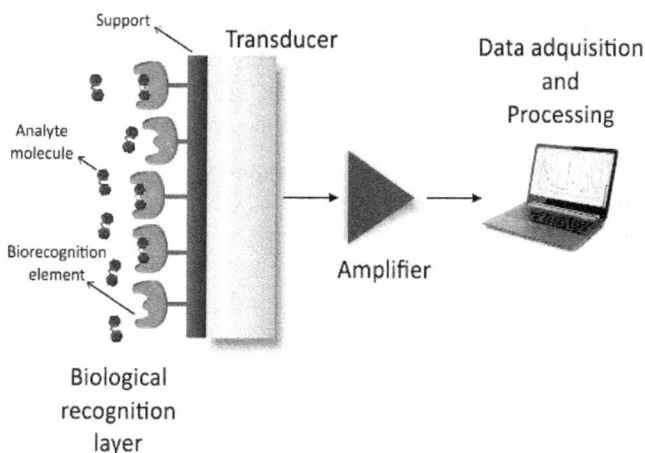

Figure 2. A typical biosensor set-up

Electrochemical biosensors can be classified according to the mode of signal transduction, such as amperometry, potentiometry or conductimetry (D'Orazio, 2003; Grieshaber et al., 2008; Mehrvar & Abdi, 2004). Amperometric approach is based on the measurement of the current resulting from the electrochemical redox reaction of electroactive species (Borgmann et al., 2011). Potentiometric biosensors involve the determination of the potential difference between two electrodes separated by a permselective membrane when zero current flows between them (Koncki, 2007). Conductometry measures changes in the ionic composition of a tested sample as a result of a biochemical reaction (Jaffrezic-Renault & Dzyadevych, 2008).

The bioreceptor and the sensor elements can be coupled together with several methods, such as, physical adsorption, entrapment, cross-linking and covalent bonding. These techniques have been summarized in many reviews (D'Souza, 2001; Sassolas, Blum & Leca-Bouvier, 2012).

4. Laccase biosensors

Laccase (p-diphenol: dioxygenoxidoreductases EC 1.10.3.2) catalyzes the oxidation of ortho- and para-diphenols, aminophenols, polyphenols, polyamines, lignins and aryl diamines with the concomitant reduction of molecular oxygen to water (Majeau, Brar & Tyagi, 2010). They are secreted by many fungi and detected in numerous plants and bacteria (Dwivedi, Singh, Pandey & Kumar, 2011). Typical fungal laccases are extracellular proteins of approximately 60-70 kDa. Laccase active site contains four copper atoms arranged in one mononuclear Type 1 site and one trinuclear cluster Type 2/Type 3. The substrates are oxidized by the Type 1 copper and the extracted electrons are transferred to the Type 2/Type 3 centre where dioxygen binds and is reduced to water (Duran, Rosa, D'Annibale & Gianfreda, 2002) (Figure 3).

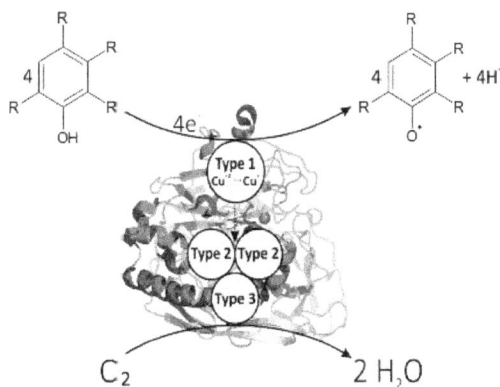

Figure 3. Catalytic cycle of laccase (Baldrian, 2006)

Laccase biosensors have been mainly focused on food industry (Chawla, Rawal, Kumar & Pundir, 2012; Di Fusco, Tortolini, Deriu & Mazzei, 2010; Gamella, Campuzano, Reviejo & Pingarrón, 2006; Martinez-Periñan, Hernández-Artiga, Palacios-Santander, ElKaoutit, Naranjo-Rodriguez & Bellido-Milla, 2011), biomedical analysis (Moccelini, Franzoi, Vieira, Dupont & Scheeren, 2011; Odaci, Timur, Pazarlioglu, Montereali, Vastarella, Pilloton et al., 2007; Shervedani & Amini, 2012) and environmental monitoring (Jia, Zhang, Wang & Wang, 2012; Oliveira, Fátima Barroso, Morais, de Lima-Neto, Correia, Oliveira et al., 2013; Wang, Tang, Zhang, Gao & Chen, 2012). Among these applications, amperometry is the most common transducer method (Liu, Qu, Guo, Chen, Liu & Dong, 2006; Rawal, Chawla, Devender & Pundir, 2012); differential pulse voltammetry (DPV) (Shervedani & Amini, 2012) and square wave voltammetry (SWV) (Moccelini et al., 2011) have been used in less extent.

A biosensor for medical applications using DPV as a transducer method was proposed by (Shervedani & Amini, 2012). *Agaricusbisporus* laccase was immobilized on 3-mercaptopropionic acid (MPA) SAM. The amine groups of the laccase were couple to the acidic groups of MPA by formation of amide group using carbodiimide hydrochloride and N-hydroxysuccinimide. The biosensor was tested with dopamine solutions and then using blood samples. A limit of detection (LOD) of 29 nM was estimated which is lower than other works using different enzymes (Fritzen-Garcia, Zoldan, Oliveira, Soldi, Pasa & Creczynski-Pasa, 2013; Moccelini, Fernandes & Vieira, 2008). Moreover, they studied the effect of interferences using ascorbic acid and uric acid. The results showed that a 25-fold excess of uric acid and 14-fold excess of ascorbic acid did not interfere in the determination of dopamine. They achieved a repeatability and reproducibility with a relative standard deviation (RSD) lower than 5%.

A biosensor for environmental monitoring of phenols was proposed using a Black Pearl 2000 modified glassy carbon electrode with laccase immobilized (Wang et al., 2012). The biosensor was evaluated with catechol by amperometry, and the kinetic Michaelis-Menten apparent constant (K_m^{app}) obtained was 1.79 mM. The LOD, linear range and sensitivity of the biosensor were 3 µM, 3 µM - 5.555 mM and 98.84 µA mM^{-1} cm^{-2}, respectively.

Liu and co-workers designed an amperometric biosensor by entrapping laccase into a matrix of carbon nanotubes (CNTs) and chitosan on a glassy carbon electrode (Liu et al., 2006). The biosensor performance was evaluated using 2,2'-azino-bis(3-ethylbenzothiazoline-6-sulphonic acid) (ABTS, a non-phenolic laccase substrate), catechol and O$_2$. For ABTS, K_m was 19.86 µM and a LOD of 0.23 µM (S/N = 3). K_m^{app} was ~ 8 fold lower compared with the previous results (Quan, Kim, Yoon & Shin, 2002) and LOD was ~ 2 fold lower than other laccase biosensor (Quan, Kim & Shin, 2004). For catechol, K_m^{app} was 9.43 µM and a LOD of 0.66 µM (S/N = 3). This K_m was between 21 and 413 fold lower than previous biosensors (Freire, Durán & Kubota, 2001; Mena, Carralero, González-Cortés, Yáñez-Sedeño & Pingarrón, 2005; Solná & Skládal, 2005; Wang et al., 2012) and LOD was ~ 20 fold lower than the screen-printed sensor based on *Coriolus hirsutus* laccase, horseradish peroxidase and mushroom tyrosinase immobilized on a 4-mercapto-1-butanol Au electrode (Solná & Skládal, 2005). In contrast with the laccase biosensor reported by Wang and co-workers (Wang et al., 2012), the LOD and K_m^{app} were 4.5 and 199 fold lower, respectively. These improvements may be attributed to the good conductivity and the enhancing effect on the electrocatalytic activity due to the presence of CNTs. Other study had shown a significant increase of the catalyzed reaction currents (around 6 fold) when multi-wall CNTs were added into a composite electrode with laccase (Oliveira et al., 2013). Besides, the reversibility was improved and the separation between peaks was lower.

This previous work pointed the current trend of taking advantage of the rapid development of nanotechnology. Currently, many biosensors are improved using nanomaterials, specially noble metals nanoparticles and CNTs.

5. Tyrosinase biosensors

Tyrosinases (E.C. 1.14.18.1) are copper proteins that catalyze two different oxygen-dependent enzymatic reactions: the hydroxylation of monophenols to o-diphenols (cresolase activity) and the subsequent oxidation of o-diphenols to o-quinones (catecholase activity) (Akyilmaz, Yorganci & Asav, 2010). Even though these enzymes are widely distributed in animals, plants, insects and microorganisms, purification and extraction of tyrosinase have focused on fungi and *Streotomyces spp.* (McMahon, Doyle, Brooks & O'Connor, 2007). In contrast to lacasse, tyrosinase molecular weight is significantly heterogeneous as several studies indicated a wide range of values (Halaouli, Asther, Sigoillot, Hamdi & Lomascolo, 2006).

Tyrosinase active sites consist on a coupled binuclear copper complex, both copper atoms are Type 3 as they are centrally located (Durán, Rosa, D'Annibale & Gianfreda, 2002). Tyrosinase catalytic cycle contains two sub-cycles (Figure 4): in the former, tyrosinase passes through several enzyme states (E_{deoxy}, E_{oxy}, E_{oxy-M}, and E_{met-D}) whilst achieving monophenol oxidation; and in the latter, as the o-diphenols are oxidized, the enzyme passes through five states (E_{deoxy}, E_{oxy}, E_{oxy-D}, E_{met}, and E_{met-D}) (Wang, Xu, Ye, Zhu & Chen, 2002). Hence, tyrosinase biosensors take advantage of the electrochemical reduction of quinones so as to measure the electric potential created.

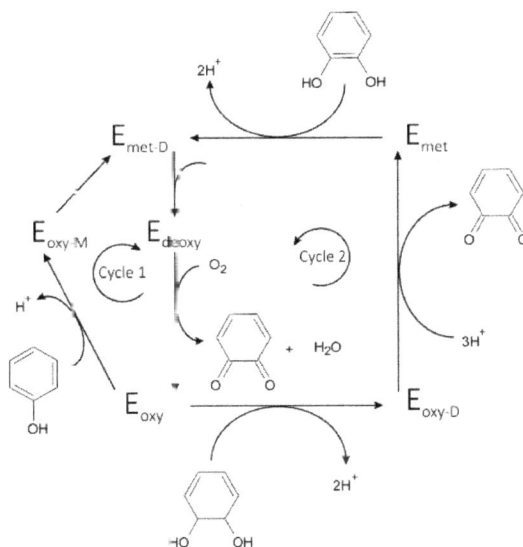

Figure 4. Catalitic cycle of tyrosinase (Seo, Sharma & Sharma, 2003)

As well as lacasse, tyrosinase has been used in biosensors principally in the food industry (Apetrei & Apetrei, 2013; Böyükbayram, Kıralp, Toppare & Yağcı, 2006; Cetó, Gutiérrez, Gutiérrez, Céspedes, Capdevila, Mínguez, et al., 2012; Ghasemi-Varnamkhasti, Rodríguez-Méndez, Mohtasebi, Apetrei Lozano, Ahmadi et al., 2012; Sánchez-Obrero, Mayén, Rodríguez-Mellado & Rodríguez-Amaro, 2012), environmental control (Li, Li, Song, Li, Zou & Long, 2012; Mayorga-Martinez, Cadevall, Guix, Ros & Merkoçi, 2013; Moczko, Istamboulie, Calas-Blanchard,

Rouillon & Noguer, 2012; Van Dyk & Pletschke, 2011; Wu, Deng, Jin, Lu & Chen, 2012), and biomedical analysis (Apetrei, Rodriguez-Mendez, Apetrei & de Saja, 2013; Bujduveanu, Yao, Le Goff, Gorgy, Shan, Diao et al., 2013; Rather, Pilehvar & De Wael, 2013). In all different applications, the most conventional transducer method for electrochemical biosensing is amperometry (Akyilmaz et al., 2010; Apetrei et al., 2013; Apetrei & Apetrei, 2013; Bujduveanu et al., 2013; Hervás Pérez, Sánchez-Paniagua López, López-Cabarcos & López-Ruiz, 2006; Mayorga-Martinez et al., 2013; Sánchez-Obrero et al., 2012; Wang et al., 2002; Wu et al., 2012). However, cyclic voltammetry (Cetó et al., 2012; Ghasemi-Varnamkhasti et al., 2012; Li et al., 2012), and conductometry (Wang, Chen, Xia, Zhu, Zhao, Chovelon et al., 2006) have been also used.

Wu and co-workers constructed a tyrosinase biosensor for the determination of bisphenol A for rapid analysis of emergency pollution affairs, using graphene as both enzyme immobilization platform and electrode material (Wu et al., 2012). The biosensor performance was assessed and compared with multi-wall CNTs modified tyrosinase biosensors, indicating that graphene-based biosensors have significant advantages in response, repeatability, background current and LOD. These biosensors showed an analytical performance over the linear range from 100 nM to 2 μM, with LOD of 33 nM and sensitivity of 3108.4 mA M^{-1} cm^{-2}.

In parallel, a disposable, low-cost and easy to carry biosensor employing single-wall CNTs, gold nanoparticles and tyrosinase modified-screen-printed electrodes for environmental phenolic analysis in water samples was developed by (Li et al., 2012). The sensor was characterized showing a high load of tyrosinase due to the large surface area of the single-wall CNTs, as well as high tyrosinase bioactivity and enhanced sensitivity. For catechol sensing, it indicated a linear range of 80 nM to 20 μM with a LOD of 45 nM (S/N = 3) and a fast response time within 10 s. In contrast, lacasse biosensors based on a carbon nanotubes and chitosan matrix have a greater LOD for catechol detection (660 nM) (Liu et al., 2006).

Yang and co-workers constructed a tyrosinase-chitosan-carbon-coated nickel nanoparticle film for catechol detection (Yang, Xiong, Zhang & Wang, 2012). When the catechol solution was added, the biosensor cathodic current reached a 95% of steady-state current within 8 s. The linear detection range was from 0.25 nM to 27 μM, the LOD (S/N = 3) was 0.083 nM, and a sensitivity of 514 μA mM^{-1}. These results displayed an improvement in biosensor performance over those based on polyaniline-ionic-liquid carbon nanofiber composite (linear range of 0.40 nM – 2.1 μM) (Zhang, Lei, Liu, Zhao & Ju, 2009), alumina sol-gel on Sonogel-carbon transducer (linear range of 0.10 – 3.0 μM) (Zejli, Hidalgo-Hidalgo de Cisneros, Naranjo-Rodriguez, Liu, Temsamani & Marty, 2008), colloidal gold nanoparticles graphite-teflon composite (linear range of 10 nM – 2.1 μM) (Carralero, Mena, Gonzalez-Cortés, Yáñez-Sedeño & Pingarrón, 2006), Zn nanorod (sensitivity of 2.14 μA mM^{-1}) (Chen, Gu, Zhu, Wu, Liu & Xu, 2008), multi-wall CNT nafion nanobiocomposites (sensitivity of 346 μA mM^{-1}) (Tsai & Chiu, 2007), and single-wall CNTs (sensitivity of 355 μA mM^{-1}) (Zhao, Guan, Gu & Zhuang, 2005).

6. Peroxidase biosensors

Peroxidases (E.C. 1.11.1) are a large family of oxidoreductases such as NADH (E.C. 1.11.1.1), glutathione peroxidase (E.C. 1.11.1.9), and catalase-peroxidase (E.C. 1.11.1.21) among others; that catalyze several organic and inorganic compounds, using hydrogen peroxide. Most of them contain iron (III) protoporphyrin IX (heme) as prosthetic group (Conesa, Punt & van den Hondel, 2002). These enzymes, which molecular weight ranges from 30 to 150 kDa, are widely distributed in vascular plants, animals and microorganisms (dos Santos Maguerroski, Fernandes, Franzoi & Vieira, 2009; Hamid & Khalil-ur, 2009).

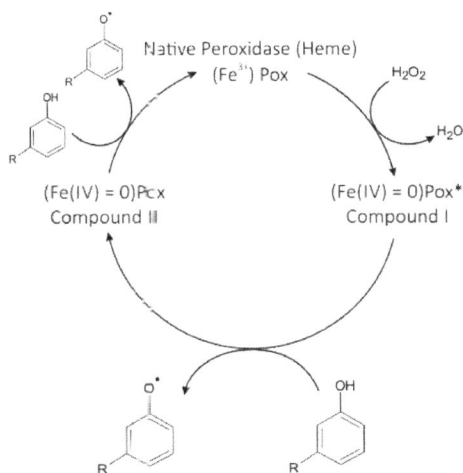

Figure 5. Catalytic cycle of peroxidase (Brill, 1966)

The classic catalytic cycle, common for most heme peroxidases, consist on the initial transfer of two H_2O_2 electrons to the enzyme state [(Fe^{3+})Pox] that leads in the formation of water and Compound I [(Fe(IV) = O)Pox*]. Then, Compound I is reduced by the oxidizing compound, in this case a phenolic compound, occurring in Compound II [(Fe(IV) = O)Pox] formation. Lastly, another one-electron transference takes place which oxidizes the oxidizing compound, resulting in the native ferric enzyme (Conesa et al., 2002; Dunford, 2010).

Peroxidases are extensively used in clinical biochemistry, aromatic chemical synthesis and peroxide removals (Hamid & Khalil-ur, 2009). In particular, horseradish peroxidase is commercially produced on large scale in clinical diagnostic kits and immunoassays (Veitch, 2004). Peroxidase-based biosensors have been also developed in environmental monitoring (Çevik, Şenel, Baykal & Abasıyanık, 2012; El Ichi, Marzouki & Korri-Youssoufi, 2009), food industry (Granero, Fernández, Agostini & Zón, 2010; Mello, Sotomayor & Kubota, 2003; Ramírez, Granero, Zón & Fernández, 2011) and biomedical analysis (Fritzen-Garcia et al., 2013; Radhapyari, Kotoky & Khan, 2013). As well as lacasse- and tyrosinase-based biosensors, the typical transducer method in peroxidase biosensors is amperometry (Çevik et al., 2012; Dai, Xu, Wu & Ju, 2005; Granero et al., 2010; Mello et al., 2003; Qiu, Chen, Wang, Li & Ma, 2013; Radhapyari et al., 2013; Ramírez et al., 2011; Rosatto, Sotomayor, Kubota & Gushikem, 2002). Nonetheless, cyclic voltammetry (Dai et al., 2005; El Ichi et al., 2009) and square wave voltammetry (Fritzen-Garcia et al., 2013) have been also used in a lower range.

Çevik and co-workers fabricated an amperometric biosensor for monitoring phenols derivates. Horseradish peroxidase was immobilized on a poly(glycidylmethacrylate) modified iron oxide nanoparticles on a gold electrode (Çevik et al., 2012). The biosensor was characterized with catechol and amperometric detection of phenols was carried out using phenol, p-crisol, 2-aminophenol and pyrogallol. Additionally, the effect of the nanoparticles was demonstrated. It was shown for different phenols that the addition of nanoparticles increased the response between four and nine times. The LOD for p-cresol, aminophenol, catechol, phenol and pyrogallol was 26, 13, 46, 28 and 48 µM respectively. The LOD for cathecol was comparatively higher than the single-wall CNTs Au nanoparticles tyrosinase biosensor on screen printed electrodes (Li et al., 2012). The sensitivity for the phenol compounds studies ranged between 1.92 and 4.37 µA mM^{-1}.

A new peroxidase named POX$_{1B}$ purified from *Allium sativum* L. was immobilized using chitosan microspheres cross-linked with glyozal on a gold electrode (El Ichi et al., 2009). The biosensor was tested using chlorophenols from the EPA's priority pollutant list and amperometry was used as a transducer method. The lowest measured concentration was 1 pM for 4-chlorophenol and pentachlorophenol and 10 nM for 2,6-dichlorophenol. The sensitivity for 6-dichlorophenol, 4-chlorophenol and pentachlorophenol were 1.5×10^6, 1.9×10^9 and 0.9×10^9 µA M^{-1}, respectively. The sensitivity was improved in contrast with a peroxidase and glucose oxidase based biosensor with an estimated sensitivity for 4-chlorophenol of 4.1×10^4 µA M^{-1} and for pentachlorophenol of 4.1×10^4 µA M^{-1} (Serra, Reviejo & Pingarrón, 2003).

A biosensor applied to dopamine determination was designed to immobilize horse radish peroxidase on dimyristoylphosphatidylcholine bilayers supported on Au (111) by dithiotreitol SAM (Fritzen-Garcia et al., 2013). Similar to the *Agaricusbisporus* laccase immobilized on MPA SAM Au electrode (Shervedani & Amini, 2012) and the carbon paste electrode with pine kernel peroxidase (Fritzen-Garcia, Oliveira, Zanetti-Ramos, Fatibello-Filho, Soldi, Pasa et al., 2009) for DA determination, a voltammetry technique was used as a transducer method. The sensor showed a linear response for dopamine concentrations from 33 µM to 1.3 mM with a LOD of 2 µM. This value was lower than the LOD using the pine kernel peroxidase (9 µM) (Fritzen-Garcia et al., 2009) but higher than the LOD using the *Agaricusbisporus* laccase biosensor (29 nM) (Shervedani & Amini, 2012).

7. Conclusion

Throughout this chapter, the principal phenolic compounds and their toxicity were overviewed. Moreover, the latest advances on lacasse-, tyrosinase-, and peroxidase-based biosensors for phenolic compound determination were described and compared. Each of these sensors are currently used and developed mainly in the food industry, environmental control, and biomedical analysis. As well as that, recently, sensor performance has been enhanced by means of the emerging science knowledge of nanotechnology.

Lacasse-based biosensors showed to be more available compared with tyrosinase and peroxidase, which are developed in less extend. Additionally, amperometry is the most conventional electrochemical transducer method in all three different enzyme biosensors because it entails higher sensitivity and allows real-time measurements.

From the reviewed biosensors, it is noticeable that enzyme-based biosensors for phenolic compound sensing can measure very low concentrations, in contrast to the conventional methods like HPLC. Hence, the development of this type of biosensor is very promising due to its several advantages.

References

Akyilmaz, E., Yorganci, E., & Asav, E. (2010). Do copper ions activate tyrosinase enzyme? A biosensor model for the solution. *Bioelectrochemistry, 78(2),* 155-160. http://dx.doi.org/10.1016/j.bioelechem.2009.09.007

Apetrei, I.M., Rodriguez-Mendez, M.L., Apetrei, C., & de Saja, J.A. (2013). Enzyme sensor based on carbon nanotubes/cobalt(II) phthalocyanine and tyrosinase used in pharmaceutical analysis. *Sensors and Actuators B: Chemical, 177(0),* 138-144. http://dx.doi.org/10.1016/j.snb.2012.10.131

Apetrei, I.M., & Apetrei, C. (2013). Amperometric biosensor based on polypyrrole and tyrosinase for the detection of tyramine in food samples. *Sensors and Actuators B: Chemical, 178(0),* 40-46. http://dx.doi.org/10.1016/j.snb.2012.12.064

ASTM International. (2012). ASTM D1783-01 Standard Test Methods for Phenolic Compounds in Water.

Baldrian, P. (2006). Fungal laccases-occurrence and properties. *Fems Microbiology Reviews, 30(2),* 215-242. http://dx.doi.org/10.1111/j.1574-4976.2005.00010.x

Borgmann, S., Schulte, A., Neugebauer, S., & Schuhmann, W. (2011). *Amperometric biosensors Advances in Electrochemical Science and Engineering.* Wiley-VCH Verlag GmbH & Co. KgaA. 1-83.

Böyükbayram, A.E., Kıralp, S., Toppare, L., & Yağcı, Y. (2006). Preparation of biosensors by immobilization of polyphenol oxidase in conducting copolymers and their use in determination of phenolic compounds in red wine. *Bioelectrochemistry, 69(2),* 164-171.

Brill, A.S. (1966). Peroxidases and catalase. *Comprehensive Biochemistry, 14,* 447-479.

Bujduveanu, M.R., Yao, W., Le Goff, A., Gorgy, K., Shan, D., Diao, G.W., et al. (2013). Multiwalled Carbon Nanotube-CaCO3 Nanoparticle Composites for the Construction of a Tyrosinase-Based Amperometric Dopamine Biosensor. *Electroanalysis, 25(3),* 613-619. http://dx.doi.org/10.1002/elan.201200245

Cammann, K. (1977). Bio-sensors based on ion-selective electrodes. *Fresenius' Zeitschrift für Analytische Chemie, 287(1),* 1-9. http://dx.doi.org/10.1007/bf00539519

Carralero, V., Mena, M.L., Gonzalez-Cortés, A., Yáñez-Sedeño, P., & Pingarrón, J.M. (2006). Development of a high analytical performance-tyrosinase biosensor based on a composite graphite–Teflon electrode modified with gold nanoparticles. *Biosensors and Bioelectronics, 22(5),* 730-736. http://dx.doi.org/10.1016/j.bios.2006.02.012

Cetó, X., Gutiérrez, J.M., Gutiérrez, M., Céspedes, F., Capdevila, J., Mínguez, S., et al. (2012). Determination of total polyphenol index in wines employing a voltammetric electronic tongue. *Analytica Chimica Acta, 732(0),* 172-179. http://dx.doi.org/10.1016/j.aca.2012.02.026

Çevik, E., Şenel, M., Baykal, A., & Abasıyanık, M.F. (2012). A novel amperometric phenol biosensor based on immobilized HRP on poly(glycidylmethacrylate)-grafted iron oxide nanoparticles for the determination of phenol derivatives. *Sensors and Actuators B: Chemical, 173(0),* 396-405. http://dx.doi.org/10.1016/j.snb.2012.07.026

Clark, Leland C., & Lyons, Champ. (1962). Electrode systems for continuous monitoring in cardiovascular surgery. *Annals of the New York Academy of Sciences, 102(1),* 29-45. http://dx.doi.org/10.1111/j.1749-6632.1962.tb13623.x

Conesa, A., Punt, P.J., & van den Hondel, C.A.M.J.J. (2002). Fungal peroxidases: molecular aspects and applications. *Journal of Biotechnology, 93(2),* 143-158. http://dx.doi.org/10.1016/S0168-1656(01)00394-7

Chawla, S., Rawal, R., Kumar, D., & Pundir, C.S. (2012). Amperometric determination of total phenolic content in wine by laccase immobilized onto silver nanoparticles/zinc oxide nanoparticles modified gold electrode. *Analytical Biochemistry, 430(1),* 16-23. http://dx.doi.org/10.1016/j.ab.2012.07.025

Chen, L., Gu, B., Zhu, G., Wu, Y., Liu, S., & Xu, C. (2008). Electron transfer properties and electrocatalytic behavior of tyrosinase on ZnO nanorod. *Journal of Electroanalytical Chemistry, 617(1),* 7-13. http://dx.doi.org/10.1016/j.jelechem.2008.01.009

D'Orazio, P. (2003). Biosensors in clinical chemistry. *Clinica Chimica Acta, 334(1-2),* 41-69. http://dx.doi.org/10.1016/S0009-8981(03)00241-9

D'Souza, S.F. (2001). Immobilization and stabilization of biomaterials for biosensor applications. *Applied Biochemistry and Biotechnology, 96(1-3),* 225-238. http://dx.doi.org/10.1385/abab:96:1-3:225

Dai, Z., Xu, X., Wu, L., & Ju, H. (2005). Detection of Trace Phenol Based on Mesoporous Silica Derived Tyrosinase-Peroxidase Biosensor. *Electroanalysis, 17(17),* 1571-1577. http://dx.doi.org/10.1002/elan.200403256

Di Fusco, M., Tortolini, C., Deriu, D., & Mazzei, F. (2010). Laccase-based biosensor for the determination of polyphenol index in wine. *Talanta, 81(1-2),* 235-240. http://dx.doi.org/10.1016/j.talanta.2009.11.063

dos Santos Maguerroski, K., Fernandes, S.C., Franzoi, A.C., & Vieira, I.C. (2009). Pine nut peroxidase immobilized on chitosan crosslinked with citrate and ionic liquid used in the construction of a biosensor. *Enzyme and Microbial Technology, 44(6-7),* 400-405. http://dx.doi.org/10.1016/j.enzmictec.2009.02.012

Dunford, H.B. (2010). *Peroxidases and Catalases: Biochemistry, Biophysics, Biotechnology and Physiology*: Wiley.

Duran, N., Rosa, M.A., D'Annibale, A., & Gianfreda, L. (2002). Applications of laccases and tyrosinases (phenoloxidases) immobilized on different supports: a review. *Enzyme and Microbial Technology, 31(7),* 907-931. http://dx.doi.org/10.1016/s0141-0229(02)00214-4

Durán, N., Rosa, M.A., D'Annibale, A., & Gianfreda, L. (2002). Applications of laccases and tyrosinases (phenoloxidases) immobilized on different supports: a review. *Enzyme and Microbial Technology, 31(7),* 907-931. http://dx.doi.org/10.1016/S0141-0229(02)00214-4

Dwivedi, U.N., Singh, P., Pandey, V.P., & Kumar, A. (2011). Structure–function relationship among bacterial, fungal and plant laccases. *Journal of Molecular Catalysis B: Enzymatic, 68(2),* 117-128. http://dx.doi.org/10.1016/j.molcatb.2010.11.002

El Ichi, S., Marzouki, M.N., & Korri-Youssoufi, H. (2009). Direct monitoring of pollutants based on an electrochemical biosensor with novel peroxidase (POX1B). *Biosensors and Bioelectronics, 24(10),* 3084-3090. http://dx.doi.org/10.1016/j.bios.2009.03.036

Freire, R.S., Durán, N., & Kubota, L.T. (2001). Effects of fungal laccase immobilization procedures for the development of a biosensor for phenol compounds. *Talanta, 54(4),* 681-686. http://dx.doi.org/10.1016/S0039-9140(01)00318-6

Fritzen-Garcia, M.B., Oliveira, I.R.W.Z., Zanetti-Ramos, B.G., Fatibello-Filho, O., Soldi, V., Pasa, A.A., et al. (2009). Carbon paste electrode modified with pine kernel peroxidase immobilized on pegylated polyurethane nanoparticles. *Sensors and Actuators B: Chemical, 139(2),* 570-575. http://dx.doi.org/10.1016/j.snb.2009.03.039

Fritzen-Garcia, M.B., Zoldan, V.C., Oliveira, I.R.W.Z., Soldi, V., Pasa, A.A., & Creczynski-Pasa, T.B. (2013). Peroxidase immobilized on phospholipid bilayers supported on au (111) by DTT self-assembled monolayers: Application to dopamine determination. *Biotechnology and Bioengineering, 110(2),* 374-382. http://dx.doi.org/10.1002/bit.24721

Gamella, M., Campuzano, S., Reviejo, A.J., & Pingarrón, J.M. (2006). Electrochemical Estimation of the Polyphenol Index in Wines Using a Laccase Biosensor. *Journal of Agricultural and Food Chemistry, 54(21),* 7960-7967. http://dx.doi.org/10.1021/jf061451r

Ghasemi-Varnamkhasti, M., Rodríguez-Méndez, M.L., Mohtasebi, S.S., Apetrei, C., Lozano, J., Ahmadi, H., et al. (2012). Monitoring the aging of beers using a bioelectronic tongue. *Food Control, 25(1),* 216-224. http://dx.doi.org/10.1016/j.foodcont.2011.10.020

Glezer, V. (2003). *Environmental Effects of Substituted Phenols The Chemistry of Phenols.* John Wiley & Sons, Ltd. 1347-1368.

Granero, A.M., Fernández, H., Agostini, E., & Zón, M.A. (2010). An amperometric biosensor based on peroxidases from Brassica napus for the determination of the total polyphenolic content in wine and tea samples. *Talanta, 83(1),* 249-255. http://dx.doi.org/10.1016/j.talanta.2010.09.016

Grieshaber, D., MacKenzie, R., Vörös, J., & Reimhult, E. (2008). Electrochemical biosensors - Sensor principles and architectures. *Sensors, 8(3),* 1400-1458.

Halaouli, S., Asther, M., Sigoillot, J.C., Hamdi, M., & Lomascolo, A. (2006). Fungal tyrosinases: new prospects in molecular characteristics, bioengineering and biotechnological applications. *Journal of Applied Microbiology, 100(2),* 219-232. http://dx.doi.org/10.1111/j.1365-2672.2006.02866.x

Hamid, M., & Khalil-ur, R. (2009). Potential applications of peroxidases. *Food Chemistry, 115(4),* 1177-1186. http://dx.doi.org/10.1016/j.foodchem.2009.02.035

Hervás Pérez, J.P., Sánchez-Paniagua López, M., López-Cabarcos, E., & López-Ruiz, B. (2006). Amperometric tyrosinase biosensor based on polyacrylamide microgels. *Biosensors and Bioelectronics, 22(3),* 429-439. http://dx.doi.org/10.1016/j.bios.2006.05.015

International Organization of Standardization. (1992). ISO 8165-1:1992 Water quality - Determination of selected monovalent phenols Gas-chromatographic method after enrichment by extraction.

International Organization of Standardization. (1999). ISO 8165-2:1999 Water quality - Determination of selected monovalent phenols Method by derivatization and gas chromatography.

Jaffrezic-Renault, N., & Dzyadevych, S. (2008). Conductometric microbiosensors for environmental monitoring. *Sensors, 8(4),* 2569-2588.

Jia, J., Zhang, S., Wang, P., & Wang, H. (2012). Degradation of high concentration 2,4-dichlorophenol by simultaneous photocatalytic—enzymatic process using TiO2/UV and laccase. *Journal of Hazardous Materials, 205-206(0),* 150-155. http://dx.doi.org/10.1016/j.jhazmat.2011.12.052

Karim, F., & Fakhruddin, A.N.M. (2012). Recent advances in the development of biosensor for phenol: a review. *Reviews in Environmental Science and Bio/Technology, 11(3)*, 261-274. http://dx.doi.org/10.1007/s11157-012-9268-9

Koncki, R. (2007). Recent developments in potentiometric biosensors for biomedical analysis. *Analytica Chimica Acta, 599(1)*, 7-15. http://dx.doi.org/10.1016/j.aca.2007.08.003

Li, Y., Li, D., Song, W., Li, M., Zou, J., & Long, Y. (2012). Rapid method for on-site determination of phenolic contaminants in water using a disposable biosensor. *Frontiers of Environmental Science & Engineering, 6(6)*, 831-838. http://dx.doi.org/10.1007/s11783-012-0393-z

Liu, Y., Qu, X.H., Guo, H.W., Chen, H.J., Liu, B.F., & Dong, S.J. (2006). Facile preparation of amperometric laccase biosensor with multifunction based on the matrix of carbon nanotubes-chitosan composite. *Biosensors and Bioelectronics, 21(12)*, 2195-2201. http://dx.doi.org/10.1016/j.bios.2005.11.014

Majeau, J.A., Brar, S.K., & Tyagi, R.D. (2010). Laccases for removal of recalcitrant and emerging pollutants. *Bioresource Technology, 101(7)*, 2331-2350. http://dx.doi.org/10.1016/j.biortech.2009.10.087

Martinez-Periñan, E., Hernández-Artiga, M.P., Palacios-Santander, J.M., ElKaoutit, M., Naranjo-Rodriguez, I., & Bellido-Milla, D. (2011). Estimation of beer stability by sulphur dioxide and polyphenol determination. Evaluation of a Laccase-Sonogel-Carbon biosensor. *Food Chemistry, 127(1)*, 234-239. http://dx.doi.org/10.1016/j.foodchem.2010.12.097

Mayorga-Martinez, C.C., Cadevall, M., Guix, M., Ros, J., & Merkoçi, A. (2013). Bismuth nanoparticles for phenolic compounds biosensing application. *Biosensors and Bioelectronics, 40(1)*, 57-62. http://dx.doi.org/10.1016/j.bios.2012.06.010

McMahon, A.M., Doyle, E.M., Brooks, S., & O'Connor, K.E. (2007). Biochemical characterisation of the coexisting tyrosinase and laccase in the soil bacterium Pseudomonas putida F6. *Enzyme and Microbial Technology, 40(5)*, 1435-1441. http://dx.doi.org/10.1016/j.enzmictec.2006.10.020

Mehrvar, M., & Abdi, M. (2004). Recent developments, characteristics, and potential applications of electrochemical biosensors. *Analytical Sciences, 20(8)*, 1113-1126.

Mello, L.D., Sotomayor, M.D.P.T., & Kubota, L.T. (2003). HRP-based amperometric biosensor for the polyphenols determination in vegetables extract. *Sensors and Actuators B: Chemical, 96(3)*, 636-645. http://dx.doi.org/10.1016/j.snb.2003.07.008

Mena, M.L., Carralero, V., González-Cortés, A., Yáñez-Sedeño, P., & Pingarrón, J.M. (2005). Laccase Biosensor Based on N-Succinimidyl-3-Thiopropionate-Functionalized Gold Electrodes. *Electroanalysis, 17(23)*, 2147-2155. http://dx.doi.org/10.1002/elan.200503345

Michałowicz, J., & Duda, W. (2007). Phenols-sources and toxicity. *Pol J Environ Stud, 16(3)*, 347-362.

Moccelini, S.K., Franzoi, A.C., Vieira, I.C., Dupont, J., & Scheeren, C.W. (2011). A novel support for laccase immobilization: Cellulose acetate modified with ionic liquid and application in biosensor for methyldopa detection. *Biosensors & Bioelectronics, 26(8)*, 3549-3554. http://dx.doi.org/10.1016/j.bios.2011.01.043

Moccelini, S.K., Fernandes, S.C.n, & Vieira, I.C. (2008). Bean sprout peroxidase biosensor based on l-cysteine self-assembled monolayer for the determination of dopamine. *Sensors and Actuators B: Chemical, 133(2)*, 364-369. http://dx.doi.org/10.1016/j.snb.2008.02.039

Moczko, E., Istamboulie, G., Calas-Blanchard, C., Rouillon, R., & Noguer, T. (2012). Biosensor employing screen-printed PEDOT:PSS for sensitive detection of phenolic compounds in water. *Journal of Polymer Science Part A: Polymer Chemistry, 50(11),* 2286-2292. http://dx.doi.org/10.1002/pola.26009

Nguyen, M.T., Kryachko, E.S., & Vanquickenborne, L.G. (2003). *General and Theoretical Aspects of Phenols The Chemistry of Phenols.* John Wiley & Sons, Ltd. 1-198.

Odaci, D., Timur, S., Pazarlioglu. N., Montereali, M.R., Vastarella, W., Pilloton, R., et al. (2007). Determination of phenolic acids using Trametes versicolor laccase. *Talanta, 71(1),* 312-317. http://dx.doi.org/10.1016/j.talanta.2006.04.032

Oliveira, T.M.B.F., Fátima Barroso, M., Morais, S., de Lima-Neto, P., Correia, A.N., Oliveira, M.B. P. P., et al. (2013). Biosensor based on multi-walled carbon nanotubes paste electrode modified with laccase for pirimicarb pesticide quantification. *Talanta, 106(0),* 137-143. http://dx.doi.org/10.1016/j.talanta.2012.12.017

Qiu, C., Chen, T., Wang, X., Li, Y., & Ma, H. (2013). Application of horseradish peroxidase modified nanostructured Au thin films for the amperometric detection of 4-chlorophenol. *Colloids and Surfaces B: Biointerfaces, 103(0),* 129-135. http://dx.doi.org/10.1016/j.colsurfb.2012.10.017

Quan, D., Kim, Y., & Shin, W. (2004). Characterization of an amperometric laccase electrode covalently immobilized on platinum surface. *Journal of Electroanalytical Chemistry, 561(1-2),* 181-189. http://dx.doi.org/10.1016/j.jelechem.2003.08.003

Quan, D., Kim, Y., Yoon, K. B., & Shin, W. (2002). Assembly of laccase over platinum oxide surface and application as an amperometric biosensor. *Bulletin of the Korean Chemical Society, 23(3),* 385-390.

Radhapyari, K., Kotoky, P., & Khan, R. (2013). Detection of anticancer drug tamoxifen using biosensor based on polyaniline probe modified with horseradish peroxidase. *Materials Science and Engineering: C, 33(2),* 583-587. http://dx.doi.org/10.1016/j.msec.2012.09.021

Ramírez, E.A., Granero, A.M., Zón, M.A., & Fernández, H. (2011). Development of an Amperometric Biosensor Based on Peroxidases from Brassica napus for the Determination of Ochratoxin a Content in Peanut Samples. *J. Biosens. Bioelectron., 3,* 2.

Rather, J.A., Pilehvar, S., & De Wael, K. (2013). A biosensor fabricated by incorporation of a redox mediator into a carbon nanotube/nafion composite for tyrosinase immobilization: detection of matairesinol, an endocrine disruptor. *Analyst, 138(1),* 204-210. http://dx.doi.org/10.1039/c2an35959f

Rawal, R., Chawla, S., Devender, & Pundir, C.S. (2012). An amperometric biosensor based on laccase immobilized onto Fe3O4NPs/cMWCNT/PANI/Au electrode for determination of phenolic content in tea leaves extract. *Enzyme and Microbial Technology, 51(4),* 179-185. http://dx.doi.org/10.1016/j.enzmictec.2012.06.001

Rosatto, S.S., Sotomayor, P.T., Kubota, L.T., & Gushikem, Y. (2002). SiO2/Nb2O5 sol–gel as a support for HRP immobilization in biosensor preparation for phenol detection. *Electrochimica Acta, 47(28),* 4451-4458. http://dx.doi.org/10.1016/S0013-4686(02)00516-9

Sánchez-Obrero, G., Mayén, M., Rodríguez-Mellado, J.M., & Rodríguez-Amaro, R. (2012). New Biosensor for Phenols Compounds Based on Gold Nanoparticle-Modified PVC/TTF-TCNQ Composite Electrode. *Int. J. Electrochem. Sci., 7,* 10952-10964.

Sassolas, A., Blum, L.J., & Leca-Bouvier, B.D. (2012). Immobilization strategies to develop enzymatic biosensors. *Biotechnology Advances, 30(3)*, 489-511. http://dx.doi.org/10.1016/j.biotechadv.2011.09.003

Seo, S.Y., Sharma, Vinay K., & Sharma, N. (2003). Mushroom Tyrosinase: Recent Prospects. *Journal of Agricultural and Food Chemistry, 51(10)*, 2837-2853. http://dx.doi.org/10.1021/jf020826f

Serra, B., Reviejo, A.J., & Pingarrón, J.M. (2003). Composite Multienzyme Amperometric Biosensors for an Improved Detection of Phenolic Compounds. *Electroanalysis, 15(22)*, 1737-1744. http://dx.doi.org/10.1002/elan.200302765

Services U.S. Department of Health and Human (2008). Toxicological Profile for Phenol. Atlanta: Agency for Toxic Substances and Disease Registry.

Shervedani, R.K., & Amini, A. (2012). Direct electrochemistry of dopamine on gold-Agaricus bisporus laccase enzyme electrode: Characterization and quantitative detection. *Bioelectrochemistry, 84*, 25-31. http://dx.doi.org/10.1016/j.bioelechem.2011.10.004

Solná, R., & Skládal, P. (2005). Amperometric Flow-Injection Determination of Phenolic Compounds Using a Biosensor with Immobilized Laccase, Peroxidase and Tyrosinase. *Electroanalysis, 17(23)*, 2137-2146. http://dx.doi.org/10.1002/elan.200403343

Thévenot, D.R., Toth, K., Durst, R.A., & Wilson, G.S. (2001). Electrochemical biosensors: recommended definitions and classification. *Biosensors and Bioelectronics, 16(1-2)*, 121-131. http://dx.doi.org/10.1016/S0956-5663(01)00115-4

Tsai, Y.C., & Chiu, C.C. (2007). Amperometric biosensors based on multiwalled carbon nanotube-Nafion-tyrosinase nanobiocomposites for the determination of phenolic compounds. *Sensors and Actuators B: Chemical, 125*(1), 10-16. http://dx.doi.org/10.1016/j.snb.2007.01.032

U.S. Environmental Protection Agency. (1993). *Method 420.4 Determination of Total Recoverable Phenolics by Semi-Automated Colorimetry*. Cincinnati, Ohio: Office of Research and Development.

U.S. Environmental Protection Agency. (1996a). *Method 604 Phenols.* Washington DC.

U.S. Environmental Protection Agency. (1996b). *Method 625 Base/Neutrals and Acids.* Washington DC.

U.S. Environmental Protection Agency. (2000). *Method 528 Determination of Phenols in Drinking Water by Solid Phase Extraction and Capillary Column Gas Chromatography/Mass Spectrometry (GC/MS)*. Cincinnati: Office of Research and Development.

Van Dyk, J.S., & Pletschke, B. (2011). Review on the use of enzymes for the detection of organochlorine, organophosphate and carbamate pesticides in the environment. *Chemosphere, 82(3)*, 291-307. http://dx.doi.org/10.1016/j.chemosphere.2010.10.033

Veitch, N.C. (2004). Horseradish peroxidase: a modern view of a classic enzyme. *Phytochemistry, 65(3)*, 249-259. http://dx.doi.org/10.1016/j.phytochem.2003.10.022

Wang, G., Xu, J.J., Ye, L.H., Zhu, J.J., & Chen, H.Y. (2002). Highly sensitive sensors based on the immobilization of tyrosinase in chitosan. *Bioelectrochemistry, 57(1)*, 33-38. http://dx.doi.org/10.1016/S1567-5394(01)00174-8

Wang, K., Tang, J., Zhang, Z., Gao, Y., & Chen, G. (2012). Laccase on Black Pearl 2000 modified glassy carbon electrode: Characterization of direct electron transfer and biological sensing properties for pyrocatechol. *Electrochimica Acta, 70(0),* 112-117. http://dx.doi.org/10.1016/j.electacta.2012.03.028

Wang, X., Chen, L., Xia, S., Zhu, Z., Zhao J., Chovelon, J.M., et al. (2006). Tyrosinase biosensor based on interdigitated electrodes for herbicides determination. *Int. J. Electrochem. Sci., 1,* 55-61.

Weber, M., & Weber, M. (2010). Phenols. In L. Pilato (Ed.), *Phenolic Resins: A Century of Progress.* Springer Berlin Heidelberg. 9-23.

Wu, L., Deng, D., Jin, J., Lu, X., & Chen, _. (2012). Nanographene-based tyrosinase biosensor for rapid detection of bisphenol A. *Biosensors and Bioelectronics, 35(1),* 193-199. http://dx.doi.org/10.1016/j.bios.2012.02.045

Yang, L., Xiong, H., Zhang, X., & Wang, S. (2012). A novel tyrosinase biosensor based on chitosan-carbon-coated nickel nanocomposite film. *Bioelectrochemistry, 84(0),* 44-48. http://dx.doi.org/10.1016/j.bioelechem.2011.11.001

Zejli, H., Hidalgo-Hidalgo de Cisneros, J.L., Naranjo-Rodriguez, I., Liu, B., Temsamani, K.R., & Marty, J.L. (2008). Phenol biosensor based on Sonogel-Carbon transducer with tyrosinase alumina sol–gel immobilization. *Analytica Chimica Acta, 612(2),* 198-203. http://dx.doi.org/10.1016/j.aca.2008.02.029

Zhang, J., Lei, J., Liu, Y., Zhao, J.i, & Ju, H. (2009). Highly sensitive amperometric biosensors for phenols based on polyaniline–ionic liquid–carbon nanofiber composite. *Biosensors and Bioelectronics, 24(7),* 1858-1863. http://dx.doi.org/10.1016/j.bios.2008.09.012

Zhao, Q., Guan, L., Gu, Z., & Zhuang, Q. (2005). Determination of Phenolic Compounds Based on the Tyrosinase- Single Walled Carbon Nanotubes Sensor. *Electroanalysis, 17(1),* 85-88. http://dx.doi.org/10.1002/elan.200403123

Chapter 6

Electrochemical detection of dopamine using graphite electrodes modified with PAMAM G4.0-64OH dendrimers in synthetic cerebrospinal fluid

G. Armendariz[1], J. Manríquez[1], A. Santamaría[2], A. Herrera-Gómez[3,4], E. Bustos[1]

[1]Centro de Investigación y Desarrollo Tecnológico en Electroquímica S.C., P.O. Box 064, C.P. 76703, Pedro Escobedo, Querétaro, México.

[2]Instituto Nacional de Neurología y Neurocirugía. Insurgentes Sur No. 3877, México, D.F., C.P. 14269, México.

[3]UAM-Azcapotzalco. Distrito Federal, 02200 México.

[4]CINVESTAV-Unidad Querétaro. Querétaro, Qro. 76230, México.

garmendariz@cideteq.mx, jmanriquez@cideteq.mx, absada@yahoo.com, aherrera@qro.cinvestav.mx, ebustos@cideteq.mx

Doi: http://dx.doi.org/10.3926/oms.122

Referencing this chapter

Armendariz, G., Manríquez, J., Santamaría, A., Herrera-Gómez, A., & Bustos, E. (2014). Electrochemical detection of dopamine using graphite electrodes modified with PAMAM G4.0-64 OH dendrimers in synthetic cerebrospinal fluid. In M. Stoytcheva & J.F. Osma (Eds.). *Biosensors: Recent Advances and Mathematical Challenges*. Barcelona: España, OmniaScience. pp. 129-140.

1. Introduction

Parkinson's disease (PD) is a degenerative disease of the central nervous system which is characterized by the trembling of the arms and legs, stiffness and rigidity of the muscles, and slow movements. PD is one of many diseases related to changes in dopamine (DA) levels, as a deficiency of this neurotransmitter in the basal ganglia is known to play a critical role in this disorder (Ezquerro, 2008; Stella & Rajewski, 1997). DA acts as a messenger in different areas of the brain for coordination of body movements. Accordingly, the main pathological symptom of PD is the selective loss of dopaminergic neurons in the substantia nigra (Stella & Rajewski, 1997). Therefore, new and improved methods for the detection and quantification of DA, which should be immediately useful for basic research and quickly adaptable for clinical purposes, are needed to better understand the pathology and possible treatments of this disease.

Currently, methods based on fluorimetry (Venton & Wightman, 2003; Lin, Qiu, Yang, Cao & Jin, 2006), radioenzymatic assays (Wassell, 1999) and chromatography (Lin et al., 2006, Bergquist, Sciubisz, Kaczor & Silberring, 2002; Peaston & Weinkove, 2004), among others, are commonly used to assess DA. However, electrochemical sensors can also quickly and inexpensively detect DA in CSF, both *in vitro* and *in vivo*. Recent articles have used different electrodes for *in vivo* measurements, although most of them employing carbon-like materials (Venton & Wightman, 2003; Lin et al., 2006; Yavich & Tiihonen, 2000; De Toledo, Santos, Cavalheiro & Mazo, 2005; Qiao, Ding & Wang, 2005; Budai, Gulya, Mészáros, Hernándi & Bali, 2010). Nevertheless, electrode design and construction can still be improved with three highly desirable adjustments: (i) smaller size for easier *in vivo* use, (ii) portability, and (iii) construction of an easy-to-use device that detects and quantifies dopamine in real time.

Previous studies note that treating carbon surfaces with oxidizing agents in gas or liquid phases results in partial surface oxidation (i.e., the formation of oxygen-containing surface functional groups) (Boehm, 1994, 2002; Barton, Evans, Halliop & Macdonald, 1997). Nitric acid, sulfuric acid, and other oxidizing media, such as ozone or oxygen plasma, are also highly effective (Ajayan, Ebbesen, Ichihashi, Ijima, Tanigaki & Hiura, 1993; Kooi, Schlecht, Burghed & Kern, 2002; Hiura, Ebbesen & Tanigaki, 1995; Banerjee, Hemraj-Benny & Wong, 2005; Klein, Melechko, McKnight, Retterer, Rack, Fowlkes et al., 2008). Molecular oxygen only attacks the basal planes of the graphite at peripheries or at defective sites such as edge planes and vacancies (Radovic, 2003; Boehm, 1966; Anderson, 1975).

After treatment, several functional groups on the carbon-like surface, such as carboxyls, carboxylic anhydrides, and lactones, are generally present (Toebes, van Heeswijk, Bitter, van Dillen, de Jong, & de Jong, 2004; Ros, van Dillen, Geus & Koningsberger, 2002; Xia, Su, Birkner, Ruppel, Wang, Woell et al., 2005; Martínez, Callejas, Benito, Cochet, Seeger, Anson et al., 2003). Hydroxyl groups at the edge of graphitic planes exhibit phenolic character, while carbonyl groups, like quinones, remain isolated, or are arranged like pyrones. *Furthermore, both ether oxygens and pyrans can substitute one carbon atom at the edge, and al*dehydes can also be present on the oxidized carbon surfaces (Kundu, Wang, Wei & Muhler, 2008).

Several experimental techniques, such as acid-base titration (Boehm, 1966, 1994, 2002), infrared spectroscopy (IR) (Ros et al., 2002; Martínez et al., 2003; Fanning & Vannice, 1993; Kastner, Pichler, Kuzmany, Curran, Blau & Weldon, 1994), temperature-programmed desorption (TPD) (Ros et al., 2002; Xia et al., 2005; Haydar, Moreno-Castilla, Ferro-Garcia, Carrasco-Marin, Rivera-Utrilla, Perrard et al., 2000; Zhou, Sui, Zhu, Li, Chen, Dai et al., 2007), and X-ray photoelectron spectroscopy (XPS) (Martínez et al., 2003; Okpalugo, Papakonstantinou, Murphy, McLaughlin & Brown, 2005; LakshminAeayanan, Toghiani & Pittman, 2004; Park, MaClain, Tian, Suib & Karwacki, 1997), can identify which oxygen functional groups are present on these oxidized carbon-like surfaces. Indeed, XPS can also correctly identify and quantify oxygen-containing functional groups from a variety of carbon materials (Toebes et al., 2004; Kundu et al., 2008; Okpalugo et al., 2005; LakshminAeayanan et al., 2004).

Due to the aforementioned need for new and improved methods of DA detection, we have designed, constructed and characterized a portable electrochemical device which detects and quantifies DA in CSF by using the dendrimer-modified pencil lead graphite (G).

2. Experimental procedure

2.1. Reactives

D-glucose (99.005%), KCl (99.6%), NaF (99.9%), HCl (38% w), $K_3Fe(CN)_6$ (99%), $K_4Fe(CN)_6$ (99%) and H_2SO_4 (97.3%) were obtained from J.T.Baker. $CaCl_2*2H_2O$ (99.0%), NaCl (99.0%), NaOH (97.0%) and $MgSO_4*7H2O$ (98.0%) were obtained from Karal. NaH_2PO_4 (98.0%) was obtained from Reprofiquin, DA (99.9%) from Sigma Chemical Co. (St. Louis, MO, USA) and Generation 4.0 PAMAM dendrimers (−OH terminal groups) from Aldrich.

The 10 mM DA standard stock solution used for DA determination was prepared in phosphate buffer (pH = 7.2, i = 0.1). Synthetic cerebrospinal fluid (SCF) was prepared as follows: 124 mM NaCl, 26 mM $NaHCO_3$, 10 mM D-glucose, 5 mM KCl, 2 mM $MgSO_4*7H_2O$, 1.09 mM NaH_2PO_4, and 1.34 mM $CaCl_2*6H_2O$.

2.2. Construction of the G–PAMAM electrode

A commercial brand of pencil lead (G, ɸ = 0.5 mm, 5 mm length) was electrochemically treated in a 0.5 M H_2SO_4 solution, and 0.6 V vs. Ag/AgCl was applied for 5 min to preferentially promote quinone on the electrode surface (García, Armendáriz, Godínez, Torres, Sepúlveda-Guzmán & Bustos, 2011; Bustos, García, Díaz-Sánchez, Juaristi, Chapman & Godínez, 2007). While similar pre-treatment procedures were recently reported by Bustos and col (García et al., 2011; Bustos et al., 2007), their use of higher potentials promoted other functional groups, such as carboxylic acid (1.6 V vs Ag/AgCl) and phtalic anhydride (2.2 V vs Ag/AgCl).

The pre-treated electrode (G-PF) was rinsed with Milli-Q water. 20 µM PAMAM dendrimer G4.0-64 OH was deposited on the electrode surface by immersion in 0.1 M NaF solution, and application of 0.6 V vs. Ag/AgCl for 1 h, as previously reported (García et al., 2011; Bustos et al., 2007).

2.3. Electrochemical characterization of naked and modified electrodes

Prior to each electrochemical experiment, the electrolytic solutions were deoxygenated by bubbling through ultra-pure nitrogen (PRAXAIR, grade 5.0) for at least 10 min. In addition, all pencil lead were inserted into Trans-erpette micropipettes tips, sealed with a butane gas flame and rinsed with Milli-Q water before use.

A BAS-Zahner potentiostat was used to perform electrochemical impedance spectroscopy. In the three-electrode system, platinum wire (0.32 cm^2) was the counter electrode and Ag/AgCl (3 M

NaCl), the reference electrode. In the two-electrode system, these electrodes were replaced by a silver wire that had recently been coated with silver chloride and rinsed with water.

To estimate the relative fractional coverage of the surface by the dendrimer, EIS measurements were performed on the modified G-PAMAM electrodes. A 10 mV amplitude wave was used around the equilibrium potential of the 0.1 M KCl solution containing the $Fe(CN)_6^{3-/4-}$ redox couple and the 1 mM $K_3Fe(CN)_6$ electroactive probe. Frequencies from 0.1 Hz to 100 kHz were applied, and impedance values were recorded at 500 discrete frequencies per decade. The charge-transfer resistance, R_{CT}, and the double layer capacitance, C_{DL}, were obtained by fitting the experimental impedance data to Randle's equivalent-circuit model, included in the BAS-Zahner software. EIS was also employed to determine the electroactive area of the graphite electrode in the 0.9 N NaF solution. After the spectra were obtained, an additional capacitance (C_f) was added to a similar equivalent-circuit model, to account for a possible accumulation of fluorine ions between graphene sheets.

Once the electrochemical impedance spectra of the electrodes in NaF solution were obtained and fitted to the equivalent-circuit model mentioned before, total capacitance was determined with the specific capacitance of graphite (0.6 $\mu F\ cm^{-2}$) to calculate the electroactive area (McCreery & Bard, 1991).

On other hand, to estimate the effective fractional coverage of the dendrimer on the graphite surface, impedance analyses in the presence of an electroactive probe were performed. The resulting Nyquist plots were fitted to the equivalent circuit and the fraction of the surface blocked by adsorbed dendrimer was calculated from Equation 1 (Bustos et al., 2007; Tokuhisa, Zhao, Baker, Phan, Dermody, García et al., 1998):

$$\theta = (R_{CT} - R^0_{CT}) / R_{CT} \tag{1}$$

where R^0_{CT} is the charge-transfer resistance for the electroactive probe molecule reaction measured on the bare graphite electrode and R_{CT} the resistance on the coated electrode. This equation, employed in previous papers, assumes that the heterogenous surface is composed of a fractional area that fully blocks electron transfer (θ) and a fractional area that is completely accesible to the probe molecule ($1-\theta$) (Bustos et al., 2007; Tokuhisa et al., 1998).

2.4. Electrochemical detection of Dopamine

Amperometric and voltamperometric measurements were carried out using a BASi Epsilon potentiostat. The current intensities employed in constructing the DA amperometric calibration curves were obtained by measuring the faradaic current of the chronoamperometric measurements. First, consecutive additions of 2 μL DA stock solution were made to a 10 mL cell containing CSF; then, the faradaic current density was measured 30 seconds after running the experiment. Detection (DL) and quantification limits (QL) were calculated using the following equations:

$$DL = 10\sigma/m \tag{2}$$

$$QL = 5\sigma/m \tag{3}$$

where σ is the standard deviation and m is the slope related with the sensitivity of the calibration curve.

2.5. Spectroscopical characterization of naked and modified electrode

The electrodes were then characterized by Raman spectroscopy and X-ray photoinduced spectroscopy (XPS). Raman spectroscopy was performed using a Thermo micro-Raman spectrometer with a 780 nm laser, and XPS using a ThermoFisher-VG instrument equipped with a monochromatic Al Kα1 (1486.7 eV) X-ray source (model XR5) and a hemispherical electron analyzer with seven channeltrons (model XPS110).

3. Results and Discussion

3.1. Characterization of naked and modified graphite electrode

Using the specific capacitance of graphite (0.6 $\mu F\ cm^{-2}$), the electroactive area of graphite electrode was calculated which was around 1.57 cm^2. Later, the rugosity factor (η) of a bare graphite lead electrode was verified by varying the exposure length from 2 to 8 mm. As the expose length increased, electroactive area and total capacitance ($C_T = C_{DL} + C_f$) also increased. Using the correlated increase in C_T and in electroactive area (A_e), the following equation was derived considering the geometric area (A_g):

$$A_e = n\ A_g \tag{4}$$

Plotting the calculated electroactive area (A_e) vs the geometric area (A_g) of the electrode resulted in a linear equation with a slope of 19.85 ($A_e = 19.85A_g - 0.03$; $r^2 = 0.9897$), where the slope calculated represents η. These results show that the composition of the electrode substrate was constant in the measured interval, leading to a possible miniaturization of the electrode. The resulting fractional coverage ($\theta = 0.18$) shows that nucleophilic attachment of the dendrimer to the electrode surface is slight. The esther-like bonds that are formed between the peripheral O atoms of PAMAM G4.0-64OH dendrimers and the O atoms of phtalic anhydride on the graphite surface could explain the low coverage, with only quinone-favored groups avaliable to act as spacers.

To verify surface transformations after electrochemical pre-treating, XPS measured the composition of the graphite electrode after oxidation in sulfuric acid. XPS spectra of a bare graphite electrode were obtained for the C 1s, Ca 2p and O 1s peaks. The peaks present in the C 1s spectra of both electrodes (Figure 1) are listed in Table 1. Each peak was assigned to a chemical species by using previously reported studies (Kundu et al., 2008); the peaks for the graphite carbon bond at 284.8 eV and the phenol / ether carbon-oxygen single bond (i.e., C-O) at 285.8 eV were present. However, the peak for the ketone / quinone carbon-oxygen double bonds (i.e., C=O) at 286.7 eV and the peak for the carboxyl / carboxylic anhydride / ester carbon bound to two oxygens (i.e.; -COO) at 287.6 eV were only seen in the pre-treated graphite electrodes. Finally, a significant decrease in the C-O peak shows that the pre-treatment of the surface successfully partially oxidized hydroquinone, creating quinone functional groups.

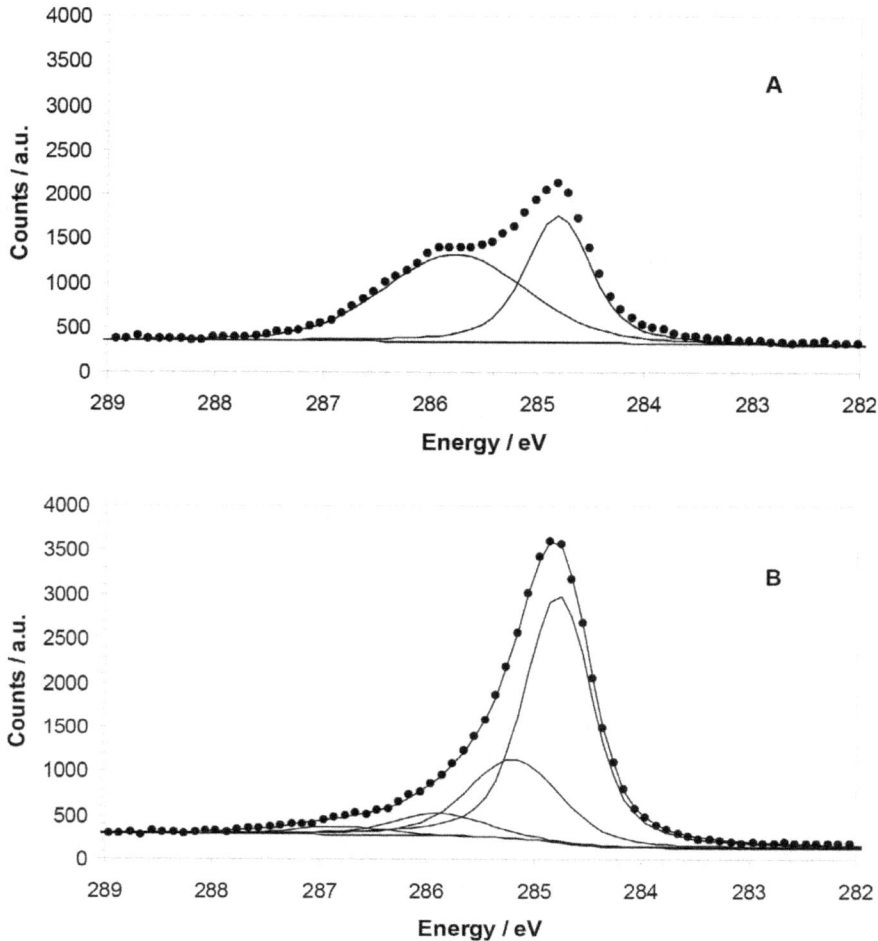

Figure 1. C 1s XPS spectra of a bare graphite electrode (A) naked and (B) pre-functionalized.
The constituting peaks corresponding to each chemical species are also shown

Electrode	Bounding Energy / eV	Assignment	Area / a.u.	Electrode	Bounding Energy / eV	Assignment	Area / a.u.
G	284.8	C=C	1259.3	G-PF	284.8	C=C	3017.52
	285.8	C-O	1876.3		285.4	C-O	717.84
	286.7	C=O	0		286.7	C=O	297.83
	287.6	-COO	0		287.6	C-O	150.34

Table 1. Relative areas determined from deconvoluted XPS spectra of bare (G) and pre-treated (G-PF)
graphite electrode

Figure 2 shows the typical Raman spectra of G (a), G-PF (b) and G-PAMAM (c). The peaks at 1360 cm^{-1} (D band) and 1580 cm^{-1} (G band) have been seen in several previous studies. The G band was assigned to the E_{2g} vibrational mode, only possible in sp^2 graphite, and the D band to the loss of symmetry along the microcrystal, like in the edges of graphene sheets (Tuinstra & Koenig, 1970). The change in the D band suggests that the number of defects in the graphite increases (from 24.50 to 41.21) when PAMAM dendrimers are coated onto the graphite surface

(Table 2). On the other hand, the increase of the G band (from 22.51 to 36.89) is most likely due to the increase of sp^2 carbon from the dendrimer branches. The lack of change in relative intensities of the D and G bands (i.e., I_D / I_G) provides evidence of edge-plane exposure. The values of I_D / I_G for all samples are similar, probably due to the low coverage of the dendrimer on the graphite surface in G-PAMAM electrodes (θ = 0.18). Furthermore, I_D / I_G ratios obtained in this study are similar to those reported by Liu (I_D / I_G = 1.4) for carbon fiber electrodes employed in the simultaneous detection of dopamine, uric acid and ascorbic acid (Liu, Huang, Hou & You, 2008).

Electrode	I_D	I_G	I_D / I_G
G	24.50	22.51	1.09
G – PF	25.33	23.00	1.10
G - PAMAM	41.21	36.89	1.12

Table 2. D (I_D) and G (I_G) band relative intensities of the electrodes, obtained from Raman spectra

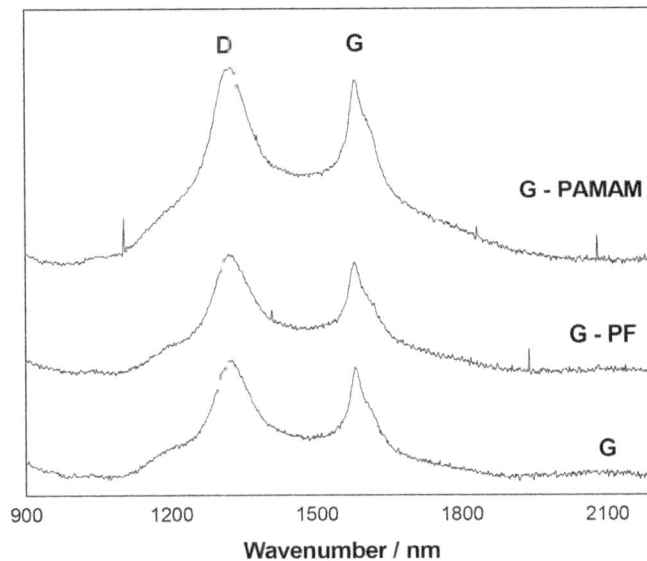

Figure 2. Raman spectra of the bare (G), the pre-treated (G-PF) and the modified electrode (G-PANAM)

3.2. Electrochemical response of dopamine using naked and modified graphite electrode

The electrochemical response of DA was examined by cyclic voltammetry (CV), using bare (A), pretreated (B), and modified (C) graphite electrodes in a three-electrode cell (Figure 3). For these experiments, a fresh 10 µM DA solution in CSF was prepared. The oxidation peak charge of DA (for the reaction DA-2e$^-$ $\rightarrow$ DOQ + 2H$^+$) obtained using G-PAMAM (1.38 µC) is much greater than those obtained using G (0.59 µC) and G-PF (0.64 µC). The increase in oxidation peak charge and the change in the shape of the voltamperogram indicates that dendrimers coated on the electrode's surface could enhance the sensitivity of the electrode. Later, the system was simplified by replacing the reference and counter electrode with a silver wire (0.32 cm^2) coated with silver chlorine. Since the composition of SCF is sufficiently high in chlorine, the electrode itself acts as an ideal non-polarizable electrode, guaranteeing accurate measurement. CV verified this asservation (Figure 4) by comparing a two-electrode (A) and three-electrode system (B) in 10μM DA solution. Though the oxidation potential of DA differed when using different electrode systems, current intensity remained constant for both. Therefore, the following experiments were carried out in a two-electrode cell.

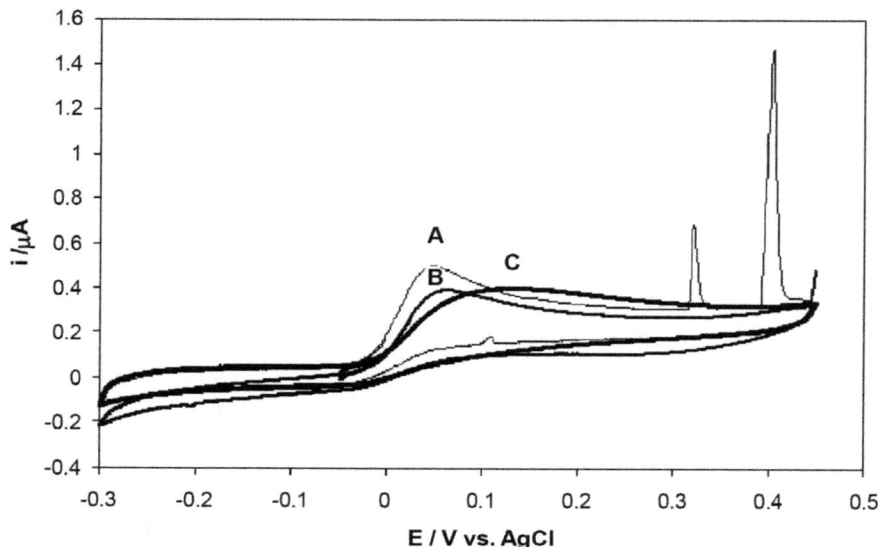

Figure 3. CVs of G (A), G – PF (B) and G – PAMAM (C) electrodes in 10 μM DA in CSF, v = 20 mV s⁻¹ using three-electrode system

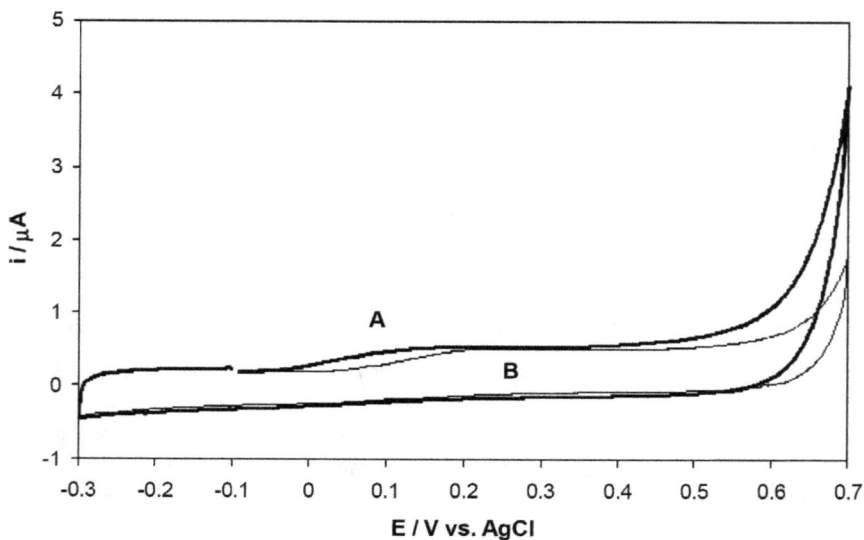

Figure 4. CV of a two (A) and a three (B) electrode system in 10 μM DA in CSF, v = 20 mV s⁻¹ using a modified electrode with G-PAMAM

3.3. Calibration curves for dopamine using naked and modified graphite electrode

A two-electrode system was employed to create calibration plots for dopamine, using DA concentrations from 0 to 20 μM. As Figure 5 shows, in this concentration range, the G-PAMAM electrode shows greater sensitivity than the bare electrode (Table 3). Although the increase in sensitivity was slight, the standard deviation of data in the absence of analyte (σ) greatly differed (by a factor of 46), greatly improving DL and QL for this electrode. In fact, according to these results, it is possible to detect 6.67 nM and quantify 22.24 nM of DA using a pencil lead graphite electrode modified with PAMAM G4.0-64OH dendrimers.

Electrode	Sensitivity ($\mu A\ \mu M^{-1}$)	DL (nM)	QL (nM)	Equation (i [=] μA; C [=] μM)	r^2
G	0.0091	306.25	1020.84	i = 0.0091[C] + 0.208	0.987
G – PF	0.0101	37.50	124.99	i = 0.0101[C] + 0.115	0.995
G - PAMAM	0.0187	6.67	22.23	i = 0.0187[C] + 0.001	0.997

Table 3. Detection (DL) and quantification limits (QL) for the electrochemical detection of DA in CSF, using G, G-PF and G-PAMAM electrodes

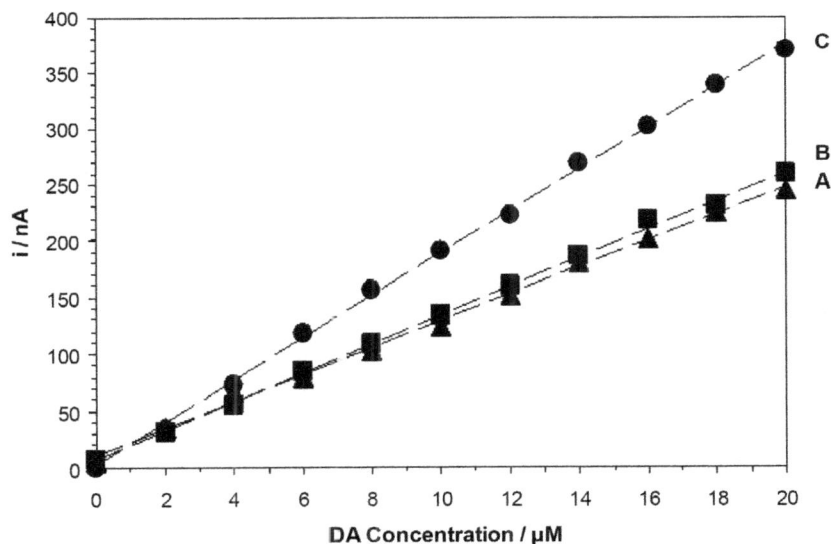

Figure 5. Amperometric calibration plots of DA in CSF using G (A), G-PF (B) and G-PAMAM (C) electrodes

4. Conclusions

Graphite pencil lead electrodes modified with PAMAM G4.0-64OH dendrimers (G-PAMAM) show good performance in detecting DA in SCF. This modified electrode was characterized by spectroscopic techniques and retained 18% of the electrode's relative surface coverage. In addition, the 3-electrode system was simplified to a 2-electrode system. Calibration plots show that the modified electrode detects and quantifies up to 6.67 nM and 22.24 nM, respectively.

Acknowledgment

The authors would like to thank the Consejo Nacional de Ciencia y Tecnología de los Estados Unidos Mexicanos (CONACyT), the Fondo Mixto del Estado de Veracruz Ignacio de la Llave (FOMIX-Veracruz), the Fondo de Cooperación Internacional de Ciencia y Tecnología Unión Europea-México (FONCICyT) ALA/2006/18149, Professor Tessy López and Prof. Yunny Meas Vöng for the funding of this research. The authors also want to thank to Lucy Yao, Peace Corps volunteer at CIDETEQ for her English revision of this manuscript.

References

Ajayan, P.M., Ebbesen, T.W., Ichihashi, T., Ijima, S., Tanigaki, T., & Hiura, H. (1993). Opening carbon nanotubes with oxygen and implications for filling. *Nature, 362*, 522-525. http://dx.doi.org/10.1038/362522a0

Anderson, J.R. (1975). *Structure of Metallic Catalysts.* U.S.: Academic Press Inc.

Banerjee, S., Hemraj-Benny, T., & Wong, S.S. (2005). Covalent surface chemistry of single-walled carbon nanotubes. *Advanced Materials, 17(1)*, 17-29. http://dx.doi.org/10.1002/adma.200401340

Barton, S.S., Evans, M.J.B., Halliop, E., & Macdonald, J.A.F. (1997). Acidic and basic sites on the surface of porous carbon. *Carbon, 35(9)*, 1361-1366. http://dx.doi.org/10.1016/S0008-6223(97)00080-8

Bergquist, J., Sciubisz, A., Kaczor, A., & Silberring, J. (2002). Catecholamines and methods for their identification and quantification in biological tissues and fluids. J*ournal of Neuroscience Methods, 113(1)*, 1-13. http://dx.doi.org/10.1016/S0165-0270(01)00502-7

Boehm, H.P. (1966). Chemical identification of surface groups. *Advanced Synthesis and Catalysis, 16*, 179-274.

Boehm, H.P. (1994). Some aspects of the surface chemistry of carbon blacks and other carbons. *Carbon, 32*, 759-769. http://dx.doi.org/10.1016/0008-6223(94)90031-0

Boehm, H.P. (2002). Surface oxides on carbon and their analysis: a critical assessment. *Carbon, 40*, 145-149. http://dx.doi.org/10.1016/S0008-6223(01)00165-8

Budai, D., Gulya K., Mészáros, B., Hernándi, I., & Bali, Z.K. (2010). Electrochemical responses of carbon fiber microelectrodes to dopamine in vitro and in vivo. *Acta Biologica Szegediensis, 54(2)*, 155-160.

Bustos, E., García, M.G., Díaz-Sánchez, B.R., Juaristi, E., Chapman, Th.W., & Godínez, L.A. (2007). Glassy carbon electrodes modified with composites of starburst-PAMAM dendrimers containing metal nanoparticles for amperometric detectionof dopamine in urine. *Talanta, 72(4)*, 1586-1592. http://dx.doi.org/10.1016/j.talanta.2007.02.017

De Toledo, R.A., Santos, M.C., Cavalheiro, E.T.G., & Mazo, L.H. (2005). Determination of dopamine in synthetic cerebrospinal fluid by SWV with a graphite-polyurethane composite electrode. *Analytical and Bioanalytical Chemistry, 381(6)*, 1161-1166. http://dx.doi.org/10.1007/s00216-005-3066-y

Ezquerro, M. (2008). Investigación en enfermedades neurodegenerativas. Evitando la epidemia "silenciosa" del siglo XXI. *Enfermería Global, 14(1)*, 1-10.

Fanning, P.E., & Vannice, M.A. (1993). A DRIFTS study of the formation of surface groups on carbon by oxidation. *Carbon, 31(5)*, 721-722. http://dx.doi.org/10.1016/0008-6223(93)90009-Y

Garcia, M.G., Armendáriz, G.M.E., Godínez, L.A., Torres, J., Sepúlveda-Guzmán, S., & Bustos, E. (2011). Detection of dopamine in non-treated urine samples using glassy carbon electrodes modified with PAMAM dendrimer-Pt composites. *Electrochimica Acta, 56(22)*, 7712-7717. http://dx.doi.org/10.1016/j.electacta.2011.06.035

Haydar, S., Moreno-Castilla, C., Ferro-Garcia, C., Carrasco-Marin, F., Rivera-Utrilla, J., Perrard, J., et al. (2000). Regularities in the temperature-programmed desorption spectra of CO2 and CO from activated carbons. *Carbon, 38(9)*, 1297-1308. http://dx.doi.org/10.1016/S0008-6223(99)00256-0

Hiura, H., Ebbesen, T.W., & Tanigaki, T. (1995). Opening and purification of carbon nanotubes in high yields. *Advanced Materials, 7(3)*, 275-276. http://dx.doi.org/10.1002/adma.19950070304

Kastner, J., Pichler, T., Kuzmany, H., Curran, S., Blau, W., & Weldon, D.N. (1994). Resonance Raman and infrared spectroscopy of carbon nanotubes. *Chemical Physics Letters, 221(1-2)*, 53-58. http://dx.doi.org/10.1016/0009-2614(94)87015-2

Klein, K.L., Melechko, A.V., McKnight, T.E., Retterer, S.T., Rack, P.D., Fowlkes, J.D., et al. (2008). Surface characterization and functionalization of carbon nanofibers. *Journal of Applied Physics, 103*, 103. http://dx.doi.org/10.1063/1.2841049

Kooi, S.E., Schlecht, U., Burghed, M., & Kern, K. (2002). Electrochemical modification of single carbon nanotubes. *Angewandte Chemie International Edition, 41(8)*, 1353-1355. http://dx.doi.org/10.1002/1521-3773(20020415)41:8<1353::AID-ANIE1353>3.0.CO;2-I

Kundu, S., Wang, Y.M., Wei, X., & Muh er, M. (2008) Thermal stability and reducibility of oxygen-containing functional groups on multiwalled carbon nanotube surfaces: a quantitative high-resolution XPS and TPD / TPR study. *The Journal of Physical Chemistry, 112(43)*, 16869-16878.

Lakshminarayanan, P.V., Toghiani, H., & Pittman, C.U. (2004). Nitric acid oxidation of vapor grown carbon nanofibers. *Carbon, 42(12-13)*, 2433-2442.
http://dx.doi.org/10.1016/j.carbon.2004.04.040

Lir, L., Qiu, P.H., Yang, L.Z., Cao, X.N., & Jin, L.T. (2006). Determination of dopamine in rat striatum by microdialysis and high-performance liquid chromatography with electrochemical detection on a functionalized multi-wall carbon nanotube electrode. *Analytical and Bioanalytical Chemistry, 384(6)*, 1308-1313. http://dx.doi.org/10.1007/s00216-005-0275-3

Liu, Y., Huang, J., Hou, H., & You, T. (2008). Simultaneous determination of dopamine, ascorbic acid and uric acid with electrospun carbon nanofibers modified electrode. *Electrochemistry Communications, 10*, 1431-1434. http://dx.doi.org/10.1016/j.elecom.2008.07.020

Martinez, M.T., Callejas, M.A., Benito, A.M., Cochet, M., Seeger, T., Anson, A., et al. (2003). Sensitivity of single wall carbon nanotubes to oxidative processing: structural modification, intercalation and functionalisation. *Carbon, 41(12)*, 2247-2256.
http://dx.doi.org/10.1016/S0008-6223(03)00250-1

McCreery, R.L., & Bard, A.J. (Eds.). (1991). *Electroanalytical Chemistry.* New York, U.S.: Dekker.

Ckpalugo, T.I.T., Papakonstantinou, P., Murphy, H., McLaughlin, J., & Brown, N.M.D. (2005). High resolution XPS characterization of chemical functionalized MWCNTs and SWCNTs. *Carbon, 43(1)*, 153-161. http://dx.doi.org/10.1016/j.carbon.2004.08.033

Park, S.H.P., MaClain, S., Tian, Z.R., Suib, S.L., & Karwacki, C. (1997). Surface and bulk measurements of metals deposited on activated carbon. *Chemistry of Materials, 9(1),* 176-183. http://dx.doi.org/10.1021/cm9602712

Peaston, R., & Weinkove, C. (2004). Measurement of catecholamines and their metabolites. *Annals of Clinical Biochemistry, 41(1),* 17-38. http://dx.doi.org/10.1258/000456304322664663

Qiao, X., Ding, H., & Wang, Z.F. (2005). Preparation of micro-biosensor and its application in monitoring in vivo change of dopamine. *Journal of Huazhong University of Science and Technology, 25(1),* 107-108. http://dx.doi.org/10.1007/BF02831402

Radovic, L.R. (2003). *Chemistry and Physics of Carbon.* New York: Marcel Dekker Inc.

Ros, T.G., van Dillen, A.J., Geus, J.W., & Koningsberger, D.C. (2002). Surface oxidation of carbon nanofibres. *Chemistry-A European Journal, 8(5),* 1151-1162. http://dx.doi.org/10.1002/1521-3765(20020301)8:5<1151::AID-CHEM1151>3.0.CO;2-#

Stella, V. J., & Rajewski, R. A. (1997). Cyclodextrins: their future in drug formulation and delivery. *Pharmaceutical Research, 14(5),* 556-557. http://dx.doi.org/10.1023/A:1012136608249

Toebes, M.L., van Heeswijk, J.M.P., Bitter, J.H., van Dillen, A.J., de Jong, & de Jong, K.P. (2004). The influence of oxidation on the texture and the number of oxygen-containing surface groups of carbon nanofibers. *Carbon, 42(2),* 307-315. http://dx.doi.org/10.1016/j.carbon.2003.10.036

Tokuhisa, H., Zhao, M., Baker, L.A., Phan, V.T., Dermody, D.L., García, M.E., et al. (1998). Preparation and characterization of dendrimer monolayers and dendrimer-alkanethiol mixed monolayers adsorbd to gold. *Journal of the American Chemical Society, 120(18),* 4492-4501. http://dx.doi.org/10.1021/ja9742904

Tuinstra, F., & Koenig, J.L. (1970). Raman spectrum of graphite. *Journal of Chemical Physics, 53(3),* 1126-1130. http://dx.doi.org/10.1063/1.1674108

Venton, B., & Wightman, M. (2003). Psychoanalytical electrochemistry: dopamine and behavior. *Analytical Chemistry, 75(19),* 414A-421A. http://dx.doi.org/10.1021/ac031421c

Wassell, J. (1999). Freedom from drug interference in new immunoassays for urinary catecholamines and metanephrines. *Clinical Chemistry, 45(12),* 2216-2223.

Xia, W., Su, D., Birkner, A., Ruppel, L., Wang, Y., Woell, Q.J., et al. (2005). Chemical vapor deposition and synthesis on carbon nanofibers: sintering of ferrocene-derived supported iron nanoparticles and the catalytic growth of secondary carbon nanofibers. *Chemistry of Materials, 17(23),* 5737-5742. http://dx.doi.org/10.1021/cm051623k

Yavich, L., & Tiihonen, J. (2000). In vivo voltammetry with removable carbon fibre electrodes in freely-moving mice: dopamine release during intracranial self-stimulation. *Journal of Neuroscience Methods, 104(1),* 53-63. http://dx.doi.org/10.1016/S0165-0270(00)00321-6

Zhou, J.H., Sui, Z.J., Zhu, J., Li, P., Chen, D., Dai, Y.C., et al. (2007). Characterization of surface oxygen complexes on carbon nanofibers by TPD, XPS and FT-IR. *Carbon, 45(4),* 785-796. http://dx.doi.org/10.1016/j.carbon.2006.11.019

Section 2

Mathematical Methods for
Biosensors Data Analysis
and Response Modeling

Chapter 7

Continuous monitoring based on biosensors coupled with artificial intelligence

Rocío B. Domínguez Cruz[1], Gustavo A. Alonso[1,2], Roberto Muñoz[1], Jean-Louis Marty[2]

[1]Department of Electrical Engineering, Bioelectronics Section, CINVESTAV, Mexico.
[2]IMAGES EA 4218, bât. S, University of Perpignan, Perpignan Cedex, France.

rdominguez@cinvestav.mx, galonso@cinvestav.mx, rmunoz@cinvestav.mx, jlmarty@univ-perp.fr

Doi: http://dx.doi.org/10.3926/oms.171

Referencing this chapter

Domínguez Cruz, R.B., Alonso, A.A., Muñoz, R. & Marty, J.L. (2014). Continuous monitoring based on biosensors coupled with artificial intelligence. In M. Stoytcheva & J.F. Osma (Eds.). *Biosensors: Recent Advances and Mathematical Challenges*. Barcelona: España, OmniaScience. pp. 143-161.

1. Introduction

Nowadays, the accurate measurement of significant parameters is an essential task for scientific and industrial fields. As examples we can mention, the monitoring of glucose for diabetic people and the environmental monitoring of emergence pollutants, such as pesticides or heavy metals, derived from industrial activity (Ogrodzki 2009, Mimendia, Gutierrez, Leija, Hernandez, Favari, Munoz et al., 2010). The evaluation of toxic compounds in food and beverage products before human consumption is an essential task as well. To carry out monitoring, specialized instrumentation such as mass spectrometry, gas or liquid chromatography and electrophoresis techniques are applied. The listed methods usually offer reliable results, ensuring that that the desired parameter is measured with high sensitivity, selectivity and accuracy even in complex samples (e.g. food samples, human blood or polluted river water) where interference can devalue the instrument response (Rodriguez-Mozaz, Lopez de Alda & Barcelo, 2007). However, a current trend in monitoring is the on-line, continuous acquisition of data from parameters of interest. Most of the mentioned methods are designed for a controlled laboratory environment rather than for an *in-situ* and continuous measurement.

As an alternative to laboratory techniques, chemical sensors were proposed as simple analytical tools providing selective information about a specific analyte in the sample. Biosensors are a special group of chemical sensors, where a biological material is used as bioreceptor for getting highly specific information about a specific analyte in the sample (Thevenot, Toth, Durst & Wilson, 1999). Nowadays, biosensors development is a growing field with thousand of publications every year devoted to environmental, food and beverage industry, security and medical applications (Ivnitski, Abdel-Hamid, Atanasov & Wilkins, 1999; Rodriguez-Mozaz, Marco, de Alda & Barcelo, 2004; Campas, Prieto-Simon & Marty, 2007; Arduini, Amine, Moscone & Palleschi, 2010). Since biosensors are base on molecular recognition, devices for specific compounds can be potentially designed for fast, low cost and portable applications such as the hand held glucose meter for diabetic people.

In spite of their widespread in scientific literature, very few biosensors have reached commercial success outside of laboratory tests. One reason could be, the single analyte approach, which is a disadvantage when compared with some powerful multianalyte established techniques (Luong, Male & Glennon, 2008). To enhance biosensor performance chemometric tools have been proposed as sensitive calibration models. Especially artificial neural networks (ANN) have been proposed as calibration tools because their ability to model non linear signals, commonly found in biosensors (Almeida, 2002). In addition, a multianalyte approach to provide information about several analytes present in sample is possible with ANN modeling (Bachmann, Leca, Vilatte, Marty, Fournier & Schmid, 2000). In spite of their advantages, chemometric tools have been barely applied to biosensors but when applied they have shown an improvement in the whole system performance.

This chapter is devoted to the review of artificial neural networks in the field of biosensors. Papers from 1995 to 2012 are presented and their applications for modeling, calibration tool for multianalyte approach is presented. Biosensing platforms applied to on-line detections in industrial process or environmental monitoring are also introduced.

2. Biosensors

2.1. Principles

In the broad sense, a sensor is a device able to convert the measurand into a measuring signal. For chemical sensors, the measurand is usually a chemical property or a specific component (analyte) in a sample of interest. The sensing information may be originated from a chemical reaction of the analyte or from the change of a physical property in the sample (Thevenot et al., 1999).

Chemical sensors have two basic sequentially connected components, namely receptor and transducer. When the receptor is based on a biochemical mechanism the whole device is called a biosensor, which is shown in figure 1. According with the International Union of Pure and Applied Chemistry (IUPAC), a biosensor is an integrated receptor-transducer device which is capable of providing selective quantitative or semi-quantitative analytical information about a specific analyte using a biological recognition element (Thevenot, Toth, Durst & Wilson, 2001). In the beginning, biosensors main applications were intended for the biomedical field (i.e. monitoring of biological samples such as glucose in human blood), but the current trends in biosensing monitoring include environmental, food, pharmaceutical and security fields (Luong et al., 2008).

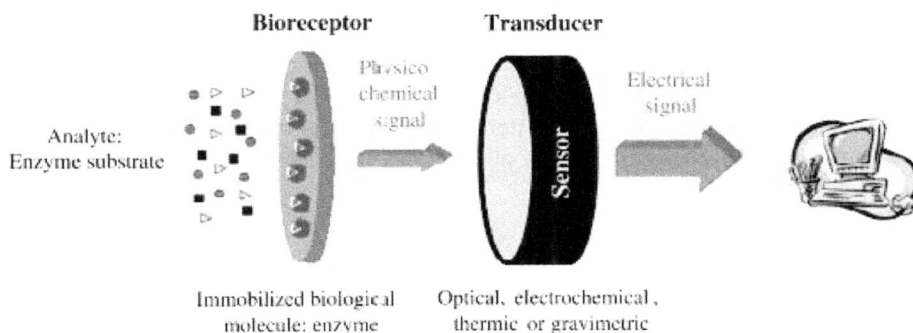

Figure 1. Typical representation of a biosensor

Biosensors can be classified according with their biological recognition method or with their signal transduction mode. For the biological recognition method two main categories are distinguished: biocatalytic devices and affinity sensors (Ronkainen, Halsall & Heineman, 2010). In biocatalytic sensors enzymes, whole cells (i.e. bacteria, fungi, eukaryotic cells) and slices of tissues (plants or animals) are used as recognition element. In these devices, the immobilized bioreceptor catalyzes a reaction which produces a detectable compound. For this group, enzymes are the most commonly found bioreceptors. The advantages of biocatalytic biosensors are the compact, easy to use and cheap design, which is exemplified in the commercially successful blood glucose biosensor (Luong et al., 2008; Chen, Xie, Yang, Xiao, Fu, Tan, et al., 2013). However, when analytes of interest are not noticeable by biocatalytic receptors or their inherent selectivity is affected by components present in complex samples, affinity biosensors are preferable. In this group the final signal is the result of the highly selective binding between the target analyte and a biomolecule (e.g. antibody). The recognition is determined by the complementary size and shape of the binding site to the analyte of interest.

In addition to bioreceptor, biosensors can be classified accordingly with their transducer mode. Electrochemical transducers cover the majority of the current reported literature on biosensors, while optical, piezoelectric and thermal transducers represent the minority (Thevenot et al., 2001). For the electrochemical transducer, the interaction between the bioreceptor and the target analyte produces detectable electroactive specie that can be measured as a current (amperometric detection) or a potential (potentiometric detection). Changes such as conductive properties in the medium or alteration on the resistance can be electrochemically detected as well. Transducers based on optical methods can transform a change in the reflectance of the surface, which is induced as a result of the interaction between the bioreceptor and the analyte. For this group, surface plasmon resonance and spectroscopy are among the most commonly found techniques in literature and in commercial available systems.

2.2. Single analyte detection

Regardless the classification, the highly selective detection of a target analyte is one of the main (if not the main) features of biosensors. In all the cases, as a result of this selective interaction a well defined signal proportional to the concentration of the target analyte is expected. For quantitative information, a mathematical relationship is established between stock solutions with well-know concentration of the studied analyte and the resulting signal. Usually, a specific representative feature of the signal (e.g. maximum current, maximum absorbance) is chosen to be related with the known concentration trough a simple linear model. Biosensor performance is tested with new samples with unknown concentration; the accuracy of determination is analytically evaluated by the coefficient of determination (R^2) and the recovery percentage. This procedure is plenty in specialized literature and suitable for controlled laboratory conditions for single target analyte detection.

2.3. Biosensor arrays

A single test is intrinsically implicit in biosensor operation since a single device can provide information about only one specific component of the analyzed sample. This can be a restriction for biosensor inclusion in analytic applications where a multianalyte approach is preferred (i.e. medical diagnosis, environmental monitoring) (Mimendia, Legin, Merkoci & del Valle, 2009; Escuder-Gilabert & Peris, 2010; Mimendia et al., 2010). For this issue sensor arrays are a good alternative. The concept of sensor array was firstly applied for gaseous sample in the analytic devices known as electronic noses (EN). Later, the concept was extended to liquid samples in the electronic tongue (ET) device. Both, EN and ET were able to provide analytical information of complex samples, while keeping a simple and relative low cost instrumentation. The main features were the inclusion of sensors with different selectivities and sensitivities towards multiple analytes and the use of high order chemometric tools for data processing (del Valle, 2010; Escuder-Gilabert & Peris, 2010). The extension of this methodology to biosensor field was possible with the inclusion of new technologies for the development of micro arrays with different immobilized bioreceptors and the inclusion of instrumentation for multichannel measurements. Besides, new immobilization methods for bioreceptors, the improving of existing biosensors and the inclusion of novel biosensors for new analytes of interest will result in wider analytical applications for biosensor arrays.

3. The role of Artificial Neural Networks in biosensor applications

3.1. Foundations of Artificial Neural Networks in biosensor data modeling

The main interest for a chemical model is to establish a relationship between a set of measurements and a set of target results (e.g. concentrations) (Rodionova & Pomerantsev, 2006). This mathematical relationship has been modeled by several ways, but in the last decades chemical field underwent a revolution with the development of sophisticated complex equipment along with high speed computer facilities, resulting in large amount of data (chemical data). Therefore, there was an urgent need to analyze such as higher order information with propel models to extract meaningful information. This fact promoted the development of the chemistry discipline known as chemometrics, which basically is dedicated to find relevant information from the measured chemical data (Rodionova & Pomerantsev, 2006). Several models (linear and non-linear) are developed, covered and studied by chemometrics. Within these models, ANNs are of great interest as modeling and calibration tools (Marini, 2009). ANNs are mathematical models attempting to mimic biological neural networks functioning in a simplified way. Even historically ANNs are related with Artificial Intelligence, they have been successfully applied in several branches of analytical chemistry (Smits, Melssen, Buydens & Kateman, 1994). History of Artificial Intelligence as discipline and ANN models has been covered by several reviews and book chapters and will not be covered here extensively. The main focus of this section is the application of ANN models for chemical science, in particularly for biosensors and biosensor arrays as alternative methods to obtain relevant information for analytical purposes.

The main ANN architectures found in chemical literature are Multilayer Perceptron (MLP) and Radial Basis Function (RBF). For MLP, usually a three-layer architecture is preferred but architectures with more layers can be found as well. When a single biosensor is modeled the input number of variables is defined by the meaningful features of biosensor measurement (e.g. peak height, peak area, maximum absorbance or intensities). When a biosensor array is modeled the number of neurons in the input layer is the number of elements in the array. Components in the output layer are usually the expected values for target analytes, but prediction of biosensor behavior can be also found as ANN output. A three layer MLP can be described as the computation of N_o functions of N_i input variables. Each function is a weighted combination of the non-linear functions computed by the neurons in the hidden layer that can be expressed as follow:

$$y_k(x) = g\left(\sum_{j=1}^{N_h} w_{kj} f\left(\sum_{i=1}^{N_I} w_{ji} x_i + w_{j0}\right) + w_{k0}\right)$$

Where

y_k is the kth component of the output vector

w_{kj} hidden to output weight

w_{ij} input to hidden weight

f and g are the non-linear activation functions

Fixing the weights in ANN architecture to find the values that best map the input experimental data to the desired output (i.e. optimization) is generally performed by the back propagation training algorithm. The goal is to change iteratively the weights between neurons in a direction that minimizes the error (E), according to the steepest descendent method (Marini, 2009).

A different architecture barely found in biosensor modeling is the Radial Basis function neural network (RBF-NN). While architectures of RBF-NN and MLP are similar, the differences are in the activation function used by RBF-NN. Radial basis function, and especially Gaussian basis function, is used in the hidden layer. Other parameters to optimize in these architectures are the center, and the scale. An output for this architecture can be written as

$$y_k(x) = \sum_{j=1}^{N_h} w_{kj}\, \rho\left(\|x - \mu_j\|\right)$$

Where

y_k is the kth component of the output vector

w_{kj} hidden to output weight

ρ radial basis function

x input vector

Preprocessing of biosensor signals is usually focused on extracting meaningful features to feed the ANN model (Jakubowska 2011). However and since biosensors are designed to be highly selective to target analyte most of the signals are univariate (i.e. steady state potential, enzyme activity). However when techniques such as cyclic voltammetry are used, preprocessing stages is necessary because of the high information order (e.g. number of records) of data (Ceto, Cespedes & del Valle, 2012). This is the case for spectroscopy records, differential pulse voltammetry and transient potentiometric records. Table 1 list the preprocessing techniques found for this review along with the input signal and target analyte.

Analyte	Signal	Preprocessing technique	Reference
Polyphenols	BA[1]	Fast Fourier Transform (FFT)	(Ceto et al., 2012)
Polyphenols	BA	Windowed sliced integral method	(Ceto, Cespedes & del Valle, 2013)
Glucose	BA	Baseline correction	(Gutes, Ibanez, del Valle & Cespedes, 2006)
Glucose, urea	SB[2]	Mean values of FIAgram segments	(Hitzmann, Ritzka, Ulber, Scheper & Schugerl, 1997)
Simulation	SB	Canonical Correlation Analysis (CCA), Principal Component Analysis (PCA)	(Baronas, Ivanauskas, Maslovskis & Vaitkus, 2004)

[1]Biosensor, [2]Single biosensor

Table 1. Preprocessing techniques for chemical data

For training the model an appropriate set of well known samples is selected from the concentration range to be modeled (Smits et al., 1994). A separate set of chemical data is usually prepared for testing the trained model. The testing set is comprised by new samples, which did not account in the training and/or by real samples (e.g. juice, wine, human blood, urine). Cross validation is the commonly applied method but some others such as jack-knife has been also reported (Marini, 2009). The accuracy of the model is measured as a function of some analytical figures of merit. The most commonly found are the sum of square errors (SSE), the root mean square error (RMSE), the relative absolute error (RAE), the coefficient of correlation (R), the coefficient of determination (R^2), the slope and intercept of the straight line formed by predicted values of ANN model *versus* the expected ones and recovery yield (Pravdova, Pravda & Guilbault, 2002; Esteban, Arino & Diaz-Cruz, 2006; Rodionova & Pomerantsev, 2006; Marini, 2009).

Finally, one of the reasons of chemometric growing is the availability of proper software for data analysis. For ANN development the table 2 list the available software and some custom implementations found in this review.

Architecture	Software	Reference
MLP	MATLAB	(Alonso, Istamboulie, Noguer, Marty & Munoz, 2012)
MLP	NEMO1.15.02	(Bachmann et al., 2000)
MLP	Turbo Pascal	(Hitzmann et al., 1997)
MLP	Stuttgart Neural Network Simulator	(Reder, Dieterle, Jansen, Alcock & Gauglitz, 2003)
MLP	Python	(Glezakos, Moschopoulou, Tsiligiridis, Kintzios & Yialouris, 2010)

Table 2. Available software for Artificial Neural Network modeling

4. Applications

This section covers the published works from 1994 to 2013 dealing with single biosensors or biosensor arrays coupled with artificial neural networks (ANN) for analytical purposes. In general ANNs are preferred as chemometric tools because their ability to accurately model non-linear data commonly found in single biosensors and biosensor arrays. The applicability of biosensors coupled with ANN is shown by the broad range fields included in this analytical methodology. Even biosensors and biosensors arrays have been modeled with other calibration tools, only those applications modeled with ANN will be considered for this section.

4.1. Modeling and simulation of biosensor response

In general, biosensors are characterized by non linear responses towards target analytes. As stated, simple calibrations do not represent accurately biosensor response in the whole range of quantification. The works in this section focus on the improvement of biosensor analytical performance towards ANN modeling. A second approach considered is the modeling of biosensor behavior after operational conditions changes. This is mainly achieved by simulation performed by ANN models and could be applied for predicting biosensor response under changing conditions

Estimation of formate with a polypyrrole based biosensor signals were performed by Talaie *et al.* (Talaie, Boger, Romagnoli, Adeloju & Yuan, 1996). Problems in the quantification were related to uncontrolled changes in the measured signals for different formate concentrations. A preprocessing was applied in order to reduce signal drifts. Baseline correction, centering and standardization were applied to the training data set. For the centered data, the highest value in electrical current and the surrounding 21 values were taken as input values for ANN model. Conjugated gradient and sigmoid transfer functions were chosen for the final model which was not validated with an external test set. In a different study, ANNs were used as multivariate calibration tools for a single glucose oxidase polypyrrole biosensor. Calibration in the direct and inverse modeling was performed to assess both, sample concentration and biosensor performance according with the measured parameters. For this fitting function task ANN obtained correlation coefficients of 0.99 for both models (direct and inverse) with an external test set (Seker & Becerik, 2004).

Hitzman *et al*. (1997) quantified glucose and urea separately from two En-FET using glucose oxidase and urease as bioreceptor. Segments of FIAgram at a fixed analyte concentration after applying different pHs were used as ANN inputs. Prediction of glucose and urea showed an average error of 4.5% and 5.5% respectively. Quantification of phenolic content was assessed from an optical tyrosinase biosensor and artificial networks. Five absorbencies intensities measured at selected wavelengths (440,490,560 and 610 nm) were the inputs of an ANN with 23 neurons in the hidden layer and 1 neuron in the output layer, corresponding to phenol concentration. Network parameters such as number of neurons in the hidden layers and training parameters such as learning rate and epochs were extensively studied and optimized to 21, 0.001 and 30000 respectively. ANN testing with additional 10 absorption spectra data showed an improvement in the range of quantification for the studied biosensor. For simple calibration linear range was limited to 0.5-6 mgL^{-1} while ANN allowed a wider quantification range of 0.5-20 mgL^{-1} (Abdullah, Ahmad, Heng, Karuppiah & Sidek, 2008).

Mixtures of ethanol and glucose were quantified in a study exploiting non-specificity of microbial biosensors (Lobanov, Borisov, Gordon, Greene, Leathers & Reshetilov, 2001). Bacterial cells of *Gluconobacter oxydans* and yeast cells of *Pischia methanolica* were used to construct amperometric biosensors. The first bioreceptor showed a high sensitivity to glucose an ethanol, while the second showed sensitivity only to ethanol. After signal preprocessing (smoothing, removal of signal peak outburst and zero drift), the rate of change of electrode current towards binary mixtures was used as biosensor response. Obtained data for binary mixtures was represented in a three dimensional space with concentrations of ethanol and glucose as abscise and ordinate and biosensor response as applicate. In a numerical approach, the resulting surfaces for both biosensors were approximated by second order polynomials. When a new (unknown concentration) sample was used, the actual concentration was determined according with the fitting in calibration surfaces for both biosensors. Alternatively, the rates of change of both biosensors along with the time from the start of measurement were used as input for an ANN model. Normalized values were used for estimation of both glucose and ethanol with a single model. Resilient back propagation was used as training algorithm and sum of squares errors (SSE) was used to assess the accuracy of analyte determination. Final coefficients of determination for polynomial approximation were 0.976 and 0.993 for glucose and ethanol respectively, while ANN showed a R^2 of 0.995 and 0.992 for the same analytes. While polynomial approximation showed a comparable performance with ANN a wider range of concentrations can be analyzed with ANN.

In a different approach, experimental signals coming from glucose and sucrose biosensors were described with ANN models (Ferreira, De Souza & Folly, 2001). The final goal was to evaluate, by simulation, the biosensor performance in a control closed- loop for an alcohol fermentation reactor. Inside the control loop biosensor response was simulated with a second order transfer function. Prediction between the control model and the biosensor response modeled by ANN were compared to study the possibility of including a biosensor in on-line control for alcohol fermentation process. Similarly in a simulation approach, synthesized data corresponding to mixtures were analyzed (Baronas et al., 2004). Data sets in batch and in flow mode were generated by numerical simulations following a full factorial design $M^K = 8^4 = 4096$ responses (M = substrate concentration, K = enzymatic rate) randomly divided in training and data set. Correlation coefficient analysis (CCA) and Principal Component Analysis (PCA) were applied to extract meaningful features from simulated data. CCA holds the point with the higher correlation for every input vector of data, while PCA retains statistically independent information from the same vector. ANN was used to distinguish between mixture components and to quantify their concentration. Recovery rates of 99% in both, batch and flow mode, were obtained.

Garcia *et al.* studied the interaction between substrate concentration, pH, and temperature in the final response of an acetylcholinesterase biosensor (Garcia, Burtseva, Stoytcheva & Gonzalez, 2011). A multilayer perceptron with 5 layers and 30 neurons in each one of the hidden layers was constructed to predict the behavior of biosensor response after a know variation of the input parameters. Experimental data produced after variations of substrate concentration (from 0.2 μmolL^{-1} to 1 μmolL^{-1}), pH (from 5 to 9) and temperature from (25°C to 70°C) were used for training and testing network with a 5 k-fold cross validation. In addition to ANN, support vector machines (SVM) were used for modeling the same data. Average MSE% for ANN prediction was found to be 2.45% in contrast to SVM with a low average value of 0.143%. Similarly, to predict the dynamic response of a potentiometric urea flow-through De Gracia *et al.* applied ANN to a single biosensor modeling (deGracia, Poch, Martorell,& Alegret, 1996) x. A four layer network with six input parameters (height at t_1 and t_2, time, injection volume, flow and concentration) and 8 neurons in the two hidden layers was used to predict peak height after changes of one of the input values. ANN prediction was compared with experimental records as well as with deterministic model performance. While deterministic model accurately described biosensor response, ANN exhibited good prediction ability after training with a reduced data set.

4.2. Environmental monitoring

Pesticide quantification is one of the most active research areas in biosensor field. Several reviews covered the trends, advances and limitations in biosensors devoted to pesticide quantification in single and multianalyte approach (Llorent-Martinez, Ortega-Barrales, Fernandez-de Cordova & Ruiz-Medina, 2011; Van Dyk & Pletschke, 2011; Pundir & Chauhan, 2012; Liu, Zheng & Li, 2013). For pesticide quantification, inhibition of acetylcholinesterase (AChE) has been used as analytical parameter in single pesticide determinations; but for pesticide mixtures, sensor arrays with engineered AChE and multivariate calibration with ANN have been proposed and successfully applied. In the first work of his kind, Bachman *et al.* used the remaining activity of AChE multisensor array (biosensor array) after pesticide exposure to quantify the presence of malaoxon and paraoxon in binary solutions (Bachmann et al., 2000). Two arrays including wild type AChE and mutant varieties (Y408F, F368L and F368H) were constructed to generate a distinctive signal pattern for sensitive multianalyte determination (Multisensor I and Multisensor II). Network architecture was optimized using magnitude based pruning and skeletonisation methods. Models with the lower RMSE were selected for ten additional runs of 3000 epochs. Multisensor II with optimized network architecture (4 input neurons, 5 neurons in the hidden layer and 1 output neuron) was applied for paraoxon quantification and network architecture of 4 input neurons, 4 neurons in the hidden layer and 1 output neuron was used for malaoxon determination in the range of 0-5 μgL^{-1}. Even both compounds exhibit similar inhibition behavior, after cross validation error prediction for paraoxon was $1.6\mu gL^{-1}$ and $0.9\mu gL^{-1}$ for malaoxon, showing the feasibility of this approach for quantification of compounds with similar behavior. A similar methodology was used by Cortina del Valle and Marty (2008) to quantify binary mixtures of dichlorvos and carbofuran mixtures Three AChEs (wild type AChE and engineered B1 and B394 varieties) were used as bioreceptors for the biosensor array. Enzyme activity after pesticide mixture exposure was spectrophotometrically studied by following Ellman method. For this assay, responses at wavelength 412 were used as ANN input, while for the electrochemical assay remaining biosensor activities after pesticide incubation were used to train the network. In both cases, data was normalized between -1 and 1 values. For each method, three layers ANNs were trained using bayesian regularization, values for learning rate and momentum were of 0.1 and 0.4 respectively. For spectrophotometric measurements the final ANN architecture had four neurons in the hidden layer with logsig function; amperometric measurements were modeled with 3 neurons in the hidden layer with tribas function. Optical method results showed a correlation coefficient of 0.995 for dichlorvos and 0.936 for carbofuran. By the other hand, electrochemical method showed correlation coefficient of0.969 for dichlorvos and 0.918 for carbofuran. Real samples evaluation for both methods le to recovery rates range of 89-116%. Following the same strategy, binary mixtures of chlorpyrifos oxon (CPO) and chlorfenvinfos (CFV) were evaluated with two different three layers ANN (Istamboulie, Cortina-Puig, Marty & Noguer, 2009). Both models had four neurons in the hidden layers, but tansig transfer function was used in the first one (ANN_1) and logsig transfer function was used in the second one (ANN_2). Correlations of 0.998 for CPO prediction and 0.995 for CFV were obtained with the first model. Correlations of 0.997 for CPO prediction and 0.986 for CFV were obtained with the second model. Spiked samples were prepared to probe the accuracy of ANN modeling resulting in recovery rates of 98% for the tested concentrations. One of the main drawbacks of methodologies based on measuring the remaining enzymatic activity after incubation of pesticide is the prolonged analysis time. Thus, an alternative method based on the slope of inhibition caused by the immediate interaction of pesticide over enzyme activity in steady state (without incubation time) was used by Alonso *et al.* for pesticide mixtures quantification (Alonso et al., 2012). In this study three electrodes based on AChE (B131 and B394 varieties) were exposed to mixtures of CPO, CFV and azinphos-methyl oxon (AZMO). The slope caused by pesticide inhibition was used to train a three layer ANN model with 3 neurons in the input layer, 95 neurons in the hidden layer and three neurons in the output layer. Values for training algorithm were studied in the range of 0.1 to 0.3 for learning rate and 0.4 to 0.9 for momentum. Network performance was evaluated according to the lower RAE. Final model resulted in a RAE of 1.82% for CPO (r = 0.985), 1.51% for CFV (r = 0.991) and 2.3% for AZMO (r = 0.997). Real samples were applied to the model for simultaneous determination of pesticide concentration. For the evaluated concentrations recovery rates were in the range of 92.05 to 105.31% for 20 evaluated samples. A method based on the spectrophotometric measurements of enzymatic kinetics was used for determination of mixtures of carbaryl and phoxim Ni *et al.* (Ni, Deng, Kokot, 2009). The signals were processed with Radial Base Function Neural Network (RBF-ANN). In addition, the highly non linear behavior of enzymatic kinetic data was processed with chemometric linear methods such as PLS2, PLS1 and PCR. The efficiency of proposed models was evaluated according to the relative prediction errors (RPE). For the linear methods RPE values were in the ranges between 8.3-15.5% for both pesticides, while for RBF-ANN model RPE value resulted in 5.2% for carbaryl and 6.5% for phoxim. The accuracy of RBF-ANN model was probed in spiked samples of lake water. Satisfactory recovery rates were obtained in the range of 98.8-103% for carbaryl and phoxim using this model.

Until now, all listed applications used biocatalytic biosensors taking advantage of the selectivity showed by AChE family towards pesticide compounds. The work proposed by Reder *et al.* employed the cross reactivity of two polyclonal antibodies to quantify the presence of two analytes: atrazine and simazine (Reder et al., 2003). Calibration experiments, dealing with the optimization of antibody presence (mixed or separated), were performed according with a full factorial design. Obtained signals were mean centered and autoscaled before ANN training. While in all cases prediction performance was lower than expected, the inclusion of different antibodies could improve the discrimination of triazines.

In a different application intended for environmental monitoring, Gutes *et al.* (2005) proposed the multivariate calibration of polyphenol oxidase amperometric biosensor with ANN for the quantification of three polyphenol compounds namely phenol, catechol and m-cresol. Synthetic mixtures of three polyphenols were electrochemically analyzed with linear sweep voltammetry. The record containing 34 measurements intensities was used to feed ANN without any preprocessing. For the ANN architecture Bayesian regularization was used as training algorithm and neuron number in the hidden layer was investigated along with a combination of transfer functions with the hidden an output layer. A combination of logsig and purelin, with five neurons in the hidden layers was chosen as optimal architecture according with the low RMSE obtained as compared with different combinations. Network predictions were near to optimal values with correlation coefficients of 0.988 for phenol, 0.997 for catechol and 0.993 for m-cresol. Even the final goal of the work was to apply the develop methodology to environmental pollution real samples analysis was not performed.

Contamination by mycrocystins mixtures (LR and YR) was assessment by modeling inhibitions of two protein phosphatase biosensors with ANN (Covaci, Sassolas, Alonso, Munoz, Radu, Bucur et al., 2012). Signals from different incubation times (20, 30 and 60 minutes) were analyzed. The network was trained using resilient back propagation, learning rate of 0.05, momentum of 0.005 and logsig transfer function. The final architecture for the network was 2 input neurons, 10 neurons in the hidden layer and two neurons in the output layer. Signals from 30 minutes of incubation time were chosen for the final model. The obtained correlation for LR was 0.996 and 0.983 for YR, while absolute errors in the range of 0,0012 to 0,0073 nM for LR and 0,012 to 0,055 nM for YR. For testing subset average recovery yields for LR was 101.73 and 105.66 for YR.

4.3. Food analysis

Besides the environmental pollution, polyphenols are present in a variety of classes as natural antioxidants. This is the case of wine and beer where polyphenols are responsible of antioxidant properties and organoleptic properties (e.g. bitterness, astringency and harshness) (Rodriguez-Mozaz et al., 2004; Ceto et al., 2012). Methods to quantify polyphenols include the Folin-Ciocalteu (FC) method, the polyphenol index (I_{280}) and High Performance Liquid Chromatography (HPLC). An alternative methodology based on a bioelectronic tongue to quantify the total polyphenol content in wine samples was proposed by Cetó *et al.* (Ceto et al., 2012). The biosensor array was comprised by four voltammetric sensors with tyrosinase or laccase enzymes as bioreceptors. The studied polyphenols were gallic acid, (±)-catechin, p-coumaric acid, caffeic acid, catechol, phenol, m-cresol, ferulic acid, chlorogenic acid and quercetin. For a first qualitative analysis, 200 µM of each of the listed polyphenols were added to different wine samples with low polyphenol content. The enhanced individual polyphenol samples were analyzed with the biosensor array and the obtained signals preprocessed using PCA. The three first components were used to represent data, with an accumulated variance of 99%. Eleven distinguished classes (10 individual polyphenols, plus wine samples without extra added polyphenol content) were classified with and ANN fed with the PCA scores. The network with 3 neurons in the input layer, 7 neurons in the hidden layer and 11 neurons with logsig function in the output layer classified the whole data. Thus sensitivity and specificity of the ANN classifier was 100% for both parameters. For a quantitative approach an ANN model was trained to predict total phenolic concentration from biosensor response after analysis of 29 wine samples of different varieties. Original biosensor array information with dimension 268 x 4 x 29 (intensities recorded x number of sensors x number of wines) was pre-processed with Fast Fourier Transform (FFT) before network training. The first 32 Fourier coefficients were used to represent each signal, leading to a compression of 88.1% as compared with the original information. Network architecture for quantification was fixed in 128 input neurons (32 coefficients for each sensor signal), 6 neurons in the hidden layers with logsig function and two output neurons with tansig function. The results provided by the ANN model were compared with established methods FC and I_{280}. The ANN methodology obtained a correlation coefficient of 0.978 when compared with FC method and of 0.949 when compared with I_{280}. The main advantage showed by this monitoring proposal is the simplicity, low cost, sensitivity and rapidity. The same group, performed and individual polyphenol analysis in beer samples following a similar methodology (Ceto et al., 2013). Basically, the same biosensor array was used to analyze synthetic mixtures of ferulic acid, gallic acid and sinapic acid. Resulting signals were pre-processed using windowed slicing integral (WSI) method. WSI divided the signals into k sections; for feature extraction the area under each section is calculated and taken as a representation (coefficient) of that specific segment of signal. For this application, biosensors signals were divided in 11 sections, and therefore represented by 11 coefficients. A network architecture of 44 input neurons, 5 neurons in the hidden layer (with tansig transfer function) and 3 neurons in the output layer provided, respectively, correlation coefficients of 0.977, 0.988 and 0.978 for ferulic acid, gallic acid and sinapic acid for testing set. Spiked real samples were applied to the system resulting in average recovery yields of 103%, 103% and 106% for ferulic, gallic and sinapic acid.

4.4. Biomedical applications

A quantification of urea and interfering alkaline ions (ammonium, potassium and sodium) with a potentiometric bioelectronic tongue in urine samples was carried out by Gutierrez et al. The biosensor array comprised all-solid state potentiometric chemosensors and biosensors modified with urease enzyme. The approach was proposed as a simple method for simultaneous quantification of urea and their common interference species in clinical samples. The data set produced by biosensor array was designed accordingly to a fractional factorial design with three levels and four factors (3^{4-1}). Calibration was performed by linear and non-linear tools (PLS1 and ANN). Network architecture of 12 input neurons (corresponding to 12 sensors in the array), 5 neurons in the hidden layers with tansig function and 4 neurons in the output layer was chosen as optimal. The network was trained with Bayesian regularization algorithm, learning rate of 0.1 and momentum of 0.4. Network architecture showed a RMSE = 0.0024 when applied to 10 additional samples used as testing set. Correlations of 0.81, 0.978, 0.995 and 0.992 were obtained for urea, ammonium, potassium and sodium respectively. The PLS1 model was tested with the same additional samples and obtained correlations of 0.727, 0.818, 0.994 and 0.919 for the same analytes. The superior performance of ANN could be attributed to the highly non-linear nature of biosensor array signals, which was accurately modeled by the non linear transfer functions in the hidden layer of ANN model. Finally, 18 urine samples were presented to both models to assess their accuracy in real clinical applications. Urine samples were previously analyzed with reference determination procedures and the obtained results compared with those computed by the models. Both, ANN and PLS1 showed a lower performance as compared with synthetic samples with slopes and intercepts away from ideal values when compared with reference methods. For ANN model this could be attributed to the high variability introduced by urine matrix in the biosensor array response, which is not considered in the training data set. Final average errors were 8% for ANN and 13% for PLS1.

The same authors extended this study adding creatinine as fifth analyte. By following the same methodology, a biosensor based on creatinine deaminase was added to the biosensor array (Gutierrez, Alegret & del Valle, 2008). Training parameters were kept but neurons in the hidden layer were fixed to 6. A superior performance as compared with the previous study was observed in the real urine samples testing. Achieved correlation coefficients were 0.967, 0.94, 0.97, 0.995 and 0.991 for urea, creatinine, ammonium, potassium and sodium respectively. However slopes and intercepts were near to ideal values when compared with reference established methods. This could be attributed to the optimization of biosensor array as compared with the first study, showing the importance of significant and meaningful input signals for an accurate modeling.

In this section some important analytes in clinical analysis such as glucose and urea were determined by biosensor array and ANN modeling. These studies probed the potential application of a simple and low cost methodology with reasonable performance for clinical samples analysis.

4.5. Agricultural applications

Identification of *Tobacco Rattle Virus* and *Cucumber Green Mottle Mosaic Virus* was determined by analyzing signals coming from BERA biosensor and ANN (Glezakos et al., 2010). BERA provided time series data with a specific pattern for each virus; since information may be affected by external factors which can affect its performance and interactive preprocessing based on genetic algorithms was applied to the time series information. The obtained data was interpreted by ANN model to identify each virus variety and success rate of 0.899 was achieved.

4.6. Pharmaceutical applications

Evaluation of signal contour by ANN was used to quantify both penicillin concentrations and potassium ion concentrations with a single enzyme field effect transistor (En-FET) coupled with flow injection analysis (Hitzmann & Kullick, 1994). Penicillin G amidase was used as bioreceptor. The effect of ion concentration (5, 15, 25, 35 and 50mM) over a fixed penicillin concentration (1.5, 2.5, 5.0, 7.5 and 10 g/L) was recorded and 7 amplitude values (from 20 to 44 seconds in steps of 4 seconds) were used as ANN inputs. For architecture, three layers networks with 3, 5, 7 and 9 neurons in the hidden layer were tested. Penicillin showed a consisted quantification with average error of 5.4% and deviations of near 10%, while potassium ion concentration showed a lower performance with an error of 3.3% but deviations as high as 38.9%.This was one of the first approaches using signal features from the same biosensor to perform multianalyte quantification.

4.7. Dedicated devices

The inclusion of automatic systems based on flow injection for liquid handling in conjunction with biosensor arrays and ANN is listed in this section. Main advantages of this approach are the full automation of the overall monitoring process, continuous operation and repeatability of the measurements and possibility of real time analysis.

Gutes *et al.* proposed the implementation of a Sequential Injection Analysis (SIA) System coupled with a voltammetric biosensor array for the determination of mixtures of glucose and ascorbic acid in fruit juices (Gutes et al., 2006). The array was composed by 3 biosensors with glucose oxidase enzyme immobilized over electrodes with different metallic catalyst (Au/Pd, Pt or Pd). Mixtures of glucose and ascorbic acid were automatically prepared by the SIA system and biosensors records (52 intensities) for each mixture were taken. Instead of using the whole records as ANN inputs record (52 intensities x 3 biosensors = 156 intensities), records from the three biosensors were added and baseline was subtracted; therefore ANN inputs were reduced to 52. A final architecture of 52 inputs, 3 neurons in the hidden layer with tansig transfer function and 2 neurons in the output layer with purelin transfer function was chosen accordingly a low RMSE of 0.4680. Correlations for this model were 0.9954 for glucose and 0.9791 for ascorbic acid in the testing test. Slopes and intercepts values were 1.015 ± 0.034 and $-2.3 \times 10^{-2} \pm 5.6 \times 10^{-2}$ for glucose while 0.915 ± 0.067 and $5.8 \times 10^{-3} \pm 5.6 \times 10^{-3}$ were obtained for ascorbic acid. After modeling, an automatic analysis with orange juice samples was performed. An acceptable performance glucose determination was obtained, with errors in a range of 0.87-12%. Determination of ascorbic acid was poorer as compared with glucose. This was attributed to the low concentration of ascorbic acid present in the samples.

A similar approach was used for the automatic determination of insecticides dichlorvos and methylparaoxon (Valdes-Ramirez, Gutierrez, del Valle, Ramirez-Silva, Fournier & Marty, 2009). As in previous studies, inhibition of AChE was used as analytical indicator of insecticide presence. In the first study, a multichannel Flow Injection System (FIA) was coupled with an enzymatic biosensor array for automatic liquid handling. Stocks of pesticides mixtures were dispensed by FIA system to the biosensor array and measurements of remaining enzymatic activity were taken. Generated data was modeled by ANN architecture of 3 input neurons, 3 neurons in the hidden layer with tansig function and 2 neurons in the output layer with purelin transfer function. Resulting slopes and intercepts from predicted values of ANN model were 0.98 ± 0.61 and 0.005 ± 0.035 for dichlorvos, while values of 0.91 ± 0.54 and 0.09 ± 0.60 were obtained for methylparaoxon. When tested with real water samples, recovery yields for dichlorvos was 104% and 118% for methylparaoxon. A similar system for *in-situ* analysis was developed by Crew *et al*. A multichannel flow system was coupled with an enzymatic biosensor array with diverse AChE varieties to obtain a distinctive inhibition pattern for organophosphate insecticides (Crew, Lonsdale, Byrd, Pittson & Hart, 2011). Calibration patterns for CFV, CPO, dichlorvos, malaoxon, chlorpyrifos methyl oxon and pyrimiphos methyl oxon were presented to the network for training. The system was applied to the detection of organophosphate insecticide presence in water and food samples by comparing the input pattern produced by biosensor array interaction with the sample with those obtained during calibration. When no significant differences from calibration were found, the absence of insecticide was assumed in both food and water. An alternative for portable instrumentation was presented by Alonso *et al*. with a hardware implementation of an ANN in a low cost chip for insecticide quantification (Alonso, Istamboulie, Ramirez-Garcia, Noguer, Marty & Munoz, 2010). A study of networks configurations (number of layers, neurons in the hidden layer, training algorithm, etc) were previously tested in MATLAB. Two networks (ANN_1 and ANN_2) were selected and implemented in a dsPIC30F6010 microcontroller. Concentrations of CPO and CFV were evaluated with the trained model. The correlation coefficients obtained for the training test with ANN_1 model were 0.992 for CPO and 0.987 for CFV with a RAE of 4.40%. For the second model (ANN_2) correlations were 0.996 for CPO and 0.994 for CFV with a RAE of 0.23%. The time for system response was 948μs ANN_1 and 800μs for ANN_2 which can be considered a real-time application.

5. Conclusions

The applications and advances of biosensor monitoring coupled with ANN in different fields were presented. First applications were mostly focused on the improvement of single biosensor performance. Usually, signal preprocessing was null and feature extraction (when utilized) was based on user experience. The main goal was improved analytical performance trough a more sophisticated calibration tool. Even good results were obtained with this approach, highly attractive applications resulted from the coupling of biosensor arrays and ANN modeling. By exploiting higher input dimensionality, cross sensitivities and non specificities in biosensors (traditionally considered as undesirable features) multianalyte detection, determination of target analytes in presence of interfering species and quantification in complex samples were achieved by using ANN modeling. These advantages has been achieved with other multivariate calibration tools; however ANN exposed a superior performance when modeling non-linearities usually found in biosensors and biosensor arrays as compared with linear models. Improvements in biosensor design, bioreceptors, transducers and measurements techniques will provide signals with higher information that could be used to improved analytical performance of the exposed approach.

References

Abdullah, J., Ahmad, M., Heng, L.Y., Karuppiah, N., & Sidek, H. (2008). Evaluation of an optical phenolic biosensor signal employing artificial neural networks. *Sensors and Actuators B-Chemical, 134(2)*, 959-965. http://dx.doi.org/10.1016/j.snb.2008.07.009

Almeida, J.S. (2002). Predictive non-linear modeling of complex data by artificial neural networks. *Current Opinion in Biotechnology, 13(1)*, 72-76. http://dx.doi.org/10.1016/S0958-1669(02)00288-4

Alonso, G.A., Istamboulie, G., Noguer, T., Marty, J.-L., & Munoz, R. (2012). Rapid determination of pesticide mixtures using disposable biosensors based on genetically modified enzymes and artificial neural networks. *Sensors and Actuators B-Chemical, 164(1)*, 22-28. http://dx.doi.org/10.1016/j.snb.2012.01.052

Alonso, G.A., Istamboulie, G., Ramirez-Garcia, A., Noguer, T., Marty, J.-L., & Munoz, R. (2010). Artificial neural network implementation in single low-cost chip for the detection of insecticides by modeling of screen-printed enzymatic sensors response. *Computers and Electronics in Agriculture, 74(2)*, 223-229. http://dx.doi.org/10.1016/j.compag.2010.08.003

Arduini, F., Amine, A., Moscone, D., & Palleschi, G. (2010). Biosensors based on cholinesterase inhibition for insecticides, nerve agents and aflatoxin B-1 detection (review). *Microchimica Acta, 170(3-4)*, 193-214. http://dx.doi.org/10.1007/s00604-010-0317-1

Bachmann, T.T., Leca, B., Vilatte, F., Marty, J.L., Fournier, D., & Schmid, R.D. (2000). Improved multianalyte detection of organophosphates and carbamates with disposable multielectrode biosensors using recombinant mutants of Drosophila acetylcholinesterase and artificial neural networks. B*iosensors & Bioelectronics, 15(3-4)*, 193-201. http://dx.doi.org/10.1016/S0956-5663(00)00055-5

Baronas, R., Ivanauskas, F., Maslovskis, R., & Vaitkus, P. (2004). An analysis of mixtures using amperometric biosensors and artificial neural networks. *Journal of Mathematical Chemistry, 36(3)*, 281-297. http://dx.doi.org/10.1023/B:JOMC.0000044225.76158.8e

Campas, M., Prieto-Simon, B., & Marty, J.L. (2007). Biosensors to detect marine toxins: Assessing seafood safety. *Talanta, 72(3)*, 884-895. http://dx.doi.org/10.1016/j.talanta.2006.12.036

Ceto, X., Cespedes, F., & del Valle, M. (2012). BioElectronic Tongue for the quantification of total polyphenol content in wine. *Talanta, 99*, 544-551. http://dx.doi.org/10.1016/j.talanta.2012.06.031

Ceto, X., Cespedes, F., & del Valle, M. (2013). Assessment of Individual Polyphenol Content in Beer by Means of a Voltammetric BioElectronic Tongue. *Electroanalysis, 25(1)*, 68-76. http://dx.doi.org/10.1002/elan.201200299

Chen, C., Xie, Q., Yang, D., Xiao, H., Fu, Y., Tan, Y., et al. (2013). Recent advances in electrochemical glucose biosensors: a review. *Rsc Advances, 3(14)*, 4473-4491. http://dx.doi.org/10.1039/c2ra22351a

Cortina, M., del Valle, M., & Marty, J.L. (2008). Electronic tongue using an enzyme inhibition biosensor array for the resolution of pesticide mixtures. *Electroanalysis, 20(1)*, 54-60. http://dx.doi.org/10.1002/elan.200704087

Covaci, O.I., Sassolas, A., Alonso, G.A., Munoz, R., Radu, G.L., Bucur, B., et al. (2012). Highly sensitive detection and discrimination of LR and YR microcystins based on protein phosphatases and an artificial neural network. *Analytical and Bioanalytical Chemistry, 404(3)*, 711-720. http://dx.doi.org/10.1007/s00216-012-6092-6

Crew, A., Lonsdale, D., Byrd, N., Pittson, R., & Hart, J.P. (2011). A screen-printed, amperometric biosensor array incorporated into a novel automated system for the simultaneous determination of organophosphate pesticides. *Biosensors & Bioelectronics, 26(6)*, 2847-2851. http://dx.doi.org/10.1016/j.bios.2010.11.018

deGracia, J., Poch, M., Martorell, D., & Alegret, S. (1996). Use of mathematical models to describe dynamic behaviour of potentiometric biosensors: Comparison of deterministic and empirical approaches to an urea flow-through biosensor. *Biosensors & Bioelectronics, 11(1-2)*, 53-61. http://dx.doi.org/10.1016/0956-5663(96)83713-4

del Valle, M. (2010). Electronic Tongues Employing Electrochemical Sensors. *Electroanalysis, 22(14)*, 1539-1555.

Escuder-Gilabert, L., & Peris, M. (2010). Review: Highlights in recent applications of electronic tongues in food analysis. *Analytica Chimica Acta, 665(1)*, 15-25. http://dx.doi.org/10.1016/j.aca.2010.03.017

Esteban, M., Arino, C., &. Diaz-Cruz, J.M (2006). Chemometrics in electroanalytical chemistry. *Critical Reviews in Analytical Chemistry, 36(3-4)*, 295-313.
http://dx.doi.org/10.1080/10408340600969381

Ferreira, L.S., De Souza, M.B., & Folly, R.O.M. (2001). Development of an alcohol fermentation control system based on biosensor measurements interpreted by neural networks. *Sensors and Actuators B-Chemical, 75(3)*, 166-171. http://dx.doi.org/10.1016/S0925-4005(01)00540-8

Garcia, E.R., Burtseva, L., Stoytcheva, M., & Gonzalez, F.F. (2011). Predicting the Behavior of the Interaction of Acetylthiocholine, pH and Temperature of an Acetylcholinesterase Sensor. Advances in Artificial Intelligence. *Proceedings 10th Mexican International Conference on Artificial Intelligence (MICAI 2011)*, 583-591.

Glezakos, T.J., Moschopoulou, G., Tsiligiridis, T.A., Kintzios, S., & Yialouris, C.P. (2010). Plant virus identification based on neural networks with evolutionary preprocessing. *Computers and Electronics in Agriculture, 70(2)*, 263-275. http://dx.doi.org/10.1016/j.compag.2009.09.007

Gutes, A., Ibanez, A.B., del Valle, M., & Cespedes, F. (2006). Automated SIA e-tongue employing a voltammetric biosensor array for the simultaneous determination of glucose and ascorbic acid. *Electroanalysis, 18(1)*, 82-88. http://dx.doi.org/10.1002/elan.200503378

Gutierrez, M., Alegret, S., & del Valle, M. (2008). Bioelectronic tongue for the simultaneous determination of urea, creatinine and alkaline ions in clinical samples. *Biosensors & Bioelectronics, 23(6)*, 795-802. http://dx.doi.org/10.1016/j.bios.2007.08.019

Hitzmann, B., & Kullick, T. (1994). Evaluation of ph field-effect transistor measurement signals by neural networks. *Analytica Chimica Acta, 294(3)*, 243-249.
http://dx.doi.org/10.1016/0003-2670(94)80307-2

Hitzmann, B., Ritzka, A., Ulber, R., Scheper, T., & Schugerl, K. (1997). Computational neural networks for the evaluation of biosensor FIA measurements. *Analytica Chimica Acta, 348(1-3)*, 135-141. http://dx.doi.org/10.1016/S0003-2670(97)00153-0

Istamboulie, G., Cortina-Puig, M., Marty, J.-L., & Noguer, T. (2009). The use of Artificial Neural Networks for the selective detection of two organophosphate insecticides: Chlorpyrifos and chlorfenvinfos. *Talanta, 79(2)*, 507-511. http://dx.doi.org/10.1016/j.talanta.2009.04.014

Ivnitski, D., Abdel-Hamid, I., Atanasov, P., & Wilkins, E. (1999). Biosensors for detection of pathogenic bacteria. *Biosensors & Bioelectronics, 14(7)*, 599-624.
http://dx.doi.org/10.1016/S0956-5663(99)00039-1

Jakubowska, M. (2011). Signal Processing in Electrochemistry. *Electroanalysis, 23(3)*, 553-572.

Liu, S., Zheng, Z., & Li, X. (2013). Advances in pesticide biosensors: current status, challenges, and future perspectives. *Analytical and Bioanalytical Chemistry, 405(1)*, 63-90.
http://dx.doi.org/10.1007/s00216-012-6299-6

Lorent-Martinez, E.J., Ortega-Barrales, P., Fernandez-de Cordova, M.L., & Ruiz-Medina, A. (2011). Trends in flow-based analytical methods applied to pesticide detection: A review. *Analytica Chimica Acta, 684(1-2)*, 30-39. http://dx.doi.org/10.1016/j.aca.2010.10.036

Lobanov, A.V., Borisov, I.A., Gordon, S.H., Greene, R.V., Leathers, T.D., & Reshetilov, A.N. (2001). Analysis of ethanol-glucose mixtures by two microbial sensors: application of chemometrics and artificial neural networks for data processing. *Biosensors & Bioelectronics, 16(9-12)*, 1001-1007.
http://dx.doi.org/10.1016/S0956-5663(01)00246-9

Luong, J.H.T., Male, K.B., & Glennon, J.D. (2008). Biosensor technology: Technology push versus market pull. *Biotechnology Advances, 26(5)*, 492-500.
http://dx.doi.org/10.1016/j.biotechadv.2008.05.007

Marini, F. (2009). Artificial neural networks in foodstuff analyses: Trends and perspectives A review. *Analytica Chimica Acta, 635(2)*, 121-131. http://dx.doi.org/10.1016/j.aca.2009.01.009

Mimendia, A., Gutierrez, J.M., Leija, L., Hernandez, P.R., Favari, L., Munoz, R., et al. (2010). A review of the use of the potentiometric electronic tongue in the monitoring of environmental systems. *Environmental Modelling & Software, 25(9)*, 1023-1030.
http://dx.doi.org/10.1016/j.envsoft.2009.12.003

Mimendia, A., Legin, A., Merkoci, A., & del Valle, M. (2009). Use of Sequential Injection Analysis to construct a Potentiometric Electronic Tongue: Application to the Multidetermination of Heavy Metals. *Olfaction and Electronic Nose, Proceedings, 1137*, 239-242.

Ni, Y., Deng, N., & Kokot, S. (2009). Simultaneous enzymatic kinetic determination of carbamate pesticides with the aid of chemometrics. *International Journal of Environmental Analytical Chemistry, 89(13)*, 939-955. http://dx.doi.org/10.1080/03067310902756151

Ogrodzki, J. (2009). Chemical sensors for water monitoring-diversity of approaches to behavioral modeling. *Proceedings of the SPIE-The International Society for Optical Engineering, 7502*, 750227.

Pravdova, V., Pravda, M., & Guilbault, G.G. (2002). Role of chemometrics for electrochemical sensors. *Analytical Letters, 35(15)*, 2389-2419. http://dx.doi.org/10.1081/AL-120016533

Pundir, C.S., & Chauhan, N. (2012). Acetylcholinesterase inhibition-based biosensors for pesticide determination: A review. *Analytical Biochemistry, 429(1)*, 19-31.
http://dx.doi.org/10.1016/j.ab.2012.06.025

Reder, S., Dieterle, F., Jansen, H., Alcock, S., &. Gauglitz, G. (2003). Multi-analyte assay for triazines using cross-reactive antibodies and neural networks. *Biosensors & Bioelectronics, 19(5)*, 447-455. http://dx.doi.org/10.1016/S0956-5663(03)00202-1

Rodionova, O.Y., & Pomerantsev, A.L. (2006). Chemometrics: Achievements and prospects. *Uspekhi Khimii, 75(4)*, 302-321. http://dx.doi.org/10.1070/RC2006v075n04ABEH003599

Rodriguez-Mozaz, S., Lopez de Alda, M.J., & Barcelo, D. (2007). Advantages and limitations of on-line solid phase extraction coupled to liquid chromatography-mass spectrometry technologies versus biosensors for monitoring of emerging contaminants in water. *Journal of Chromatography A, 1152(1-2)*, 97-115. http://dx.doi.org/10.1016/j.chroma.2007.01.046

Rodriguez-Mozaz, S., Marco, M.P., de Alda, M.J.L., & Barcelo, D. (2004). Biosensors for environmental applications: Future development trends. *Pure and Applied Chemistry, 76(4)*, 723-752. http://dx.doi.org/10.1351/pac200476040723

Ronkainen, N.J., Halsall, H.B., & Heineman, W.R. (2010). Electrochemical biosensors. *Chemical Society Reviews, 39(5)*, 1747-1763. http://dx.doi.org/10.1039/b714449k

Seker, S., & Becerik, I. (2004). A neural network model in the calibration of glucose sensor based on the immobilization of glucose oxidase into polypyrrole matrix. *Electroanalysis, 16(18)*, 1542-1549. http://dx.doi.org/10.1002/elan.200302974

Smits, J.R.M., Melssen, W.J., Buydens, L.M.C., & Kateman, G. (1994). Using artificial neural networks for solving chemical problems. 1. Multilayer feedforward networks. *Chemometrics and Intelligent Laboratory Systems, 22(2)*, 165-189. http://dx.doi.org/10.1016/0169-7439(93)E0035-3

Talaie, A., Boger, Z., Romagnoli, J.A., Adeloju, S.B., & Yuan, Y.J. (1996). Data acquisition, signal processing and modelling: A study of a conducting polypyrrole formate biosensor. 1. *Batch experiment. Synthetic Metals, 83(1)*, 21-26. http://dx.doi.org/10.1016/S0379-6779(97)80048-3

Thevenot, D.R., Toth, K., Durst, R.A., & Wilson, G.S. (1999). Electrochemical biosensors: Recommended definitions and classification - (Technical Report). *Pure and Applied Chemistry, 71(12)*, 2333-2348. http://dx.doi.org/10.1351/pac199971122333

Thevenot, D.R., Toth, K., Durst, R.A., & Wilson, G.S., (2001). Electrochemical biosensors: recommended definitions and classification. *Biosensors & Bioelectronics, 16(1-2)*, 121-131.

Valdes-Ramirez, G., Gutierrez, M., del Valle, M., Ramirez-Silva, M.T., Fournier, D., & Marty, J.L. (2009). Automated resolution of dichlorvos and methylparaoxon pesticide mixtures employing a Flow Injection system with an inhibition electronic tongue. *Biosensors & Bioelectronics, 24(5)*, 1103-1108. http://dx.doi.org/10.1016/j.bios.2008.06.022

Van Dyk, J.S., & Pletschke, B. (2011). Review on the use of enzymes for the detection of organochlorine, organophosphate and carbamate pesticides in the environment. *Chemosphere, 82(3)*, 291-307. http://dx.doi.org/10.1016/j.chemosphere.2010.10.033

Chapter 8

Modeling a second-generation glucose oxidase biosensor with statistical machine learning methods

Livier Rentería-Gutiérrez[1], Lluís A. Belanche-Muñoz[2], Félix F. González-Navarro[1*], Margarita Stilianova-Stoytcheva

[1]Instituto de Ingeniería. Blvd. Benito Juárez y Calle de la Normal S/N 21280 Mexicali, Mexico.

[2]Universitat Politècnica de Catalunya. Dept. de Llenguatges i Sistemes Informàtics C/Jordi Girona, 1-3, 08034 Barcelona, Spain.

livier.renteria@uabc.edu.mx, belanche@lsi.upc.edu, fernando.gonzalez@uabc.edu.mx, margarita.stoytcheva@uabc.edu.mx

*Corresponding author

Doi: http://dx.doi.org/10.3926/oms.165

Referencing this chapter

Rentería-Gutiérrez, L., Belanche-Muñoz, L.A., González-Navarro, F.F., & Stilianova-Stoytcheva, M. (2014). Modeling a Second-Generation Glucose Oxidase Biosensor with Statistical Machine Learning Methods. In M. Stoytcheva & J.F. Osma (Eds.). *Biosensors: Recent Advances and Mathematical Challenges*. Barcelona: España, OmniaScience. pp. 163-183.

1. Introduction

Biosensors are analytic compact devices that embody a biological piece of detection called a bio-receptor, usually formed by enzymes, microorganisms, immunoreceptors, cell receptors or chemoreceptors in current technology. They are coupled to a physical-chemical transducer that translates the biological signal to a measurable electrical signal, that is proportional to the concentration of the target compound or group of compounds to be assessed. Enzymes are mainly favored in biosensor construction because they have the capability to recognize a specific molecule (Thévenot, Toth, Durst & Wilson, 2001). One of the most attractive advantages of this sensing technology is its capacity to provide electrochemical readings in a fast, continuous and highly sensitive way. Moreover, they are susceptible to be miniaturized and its electrical response potential (or electric current) can be easily processed by cheap and compact instrumentation devices (Morrison, Dokmeci, Demirci & Khademhosseini, 2008).

The potential use of biosensors has been extended to several fields of science and engineering. Contaminant detection of the water resources (Saharudin & Asim, 2006); pathogen agent detection (Pohanka, M., Skládal & Kroca, M. (2007); drug detection in the food industry (Elliott, 2006) are only a few examples. In particular, the impact and benefits in the medical field is beyond doubt. The monitoring of lactate, urea, cholesterol or glucose are some of the body-essential features related to this technology (Department of Trade and industry, 2012; Malhotra & Chaubey, 2003). Disease monitoring and diagnosis require well-trained and qualified personnel for data acquisition and testing. It is worth mentioning that these medical tasks are highly delicate, in both sensitivity –the possibility to measure a false positive– and specificity –a false negative detection (Malhotra & Chaubey, 2003). In an ordinary scenario, a patient requires a physical examination and laboratory tests needing a few days to obtain medical results. Such a delay can sometimes derive in complications due to the lack of the proper medical treatment. The fast response, low cost and design simplicity of the biosensor approach makes it a very promising technological device in public health applications.

In the following sections, the modeling of Second-generation Electrochemical Glucose Oxidase Amperometric Biosensors (GOABs from now on) will be discussed. The importance of such devices will be contextualized in one of the most critical and prevalent diseases nowadays, the Diabetes Mellitus (DM). Machine Learning (ML) is a field within computer science and a very active area, playing an important role in science, finance and industry. It consists of a wide spectrum of methods, techniques and algorithms that aim at learning from data to find useful information or predictive models of a phenomenon. Classical and statistical ML techniques for regression are used for the modeling of the response of a GOAB. The rest of this chapter is organized as follows: In section 2, a few general ideas about DM are given, to introduce the reader into the importance and convenience to deal with the DM in its collateral consequences with this arising technology. Some general concepts about GOABs and GOAB modeling are given in section 3; section 4 describes the specific GOAB dataset and the statistical machine learning techniques used in this study. The experimental results are presented and discussed in section 6. The chapter ends with the conclusions and final thoughts.

2. Diabetes Mellitus

The DM is a serious condition where patients present high levels of blood glucose. This is due to two possible causes: a deficient insulin production or a malfunction in a particular type of cells called islets. These cells do not respond properly to insulin in the blood-glucose regulation process. In the food consumption process by humans, the body converts the inputs into glucose. A critical organ in the human body, the pancreas, produces insulin to convert this glucose into energy. When a patient is diagnosed with the DM disease, the whole process presents an erratic dynamics (Diabetes Research Institute, 2012).

There exist two types of DM: Type I includes patients that are insulin-dependent, where the islet cells are not recognized as part of the body by the Immune System and are consequently destroyed; as a result, insulin is not produced anymore; Type II embraces those diagnosed patients that produce part of their insulin needs, but not enough to maintain acceptable blood-glucose levels.

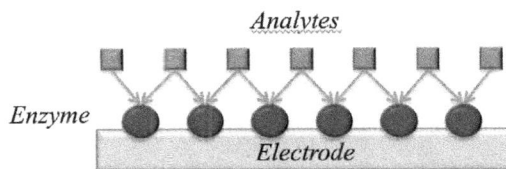

Figure 1. General scheme of an enzimatic biosensor

World Health Organization (WHO) statistics report that around 347 million of people worldwide have diabetes. Mortality estimates hits around 3.4 million deaths from consequences of high fasting blood sugar. Low- and middle-income countries have the worst mortality scenario, where more than 80% of diabetes patients die today. The WHO foresees the DM to be the 7th leading cause of death by 2030 (World Health Organization, 2013). Thus the continuous glucose monitoring by biosensors can be very helpful to diagnosed patients to prevent acute or chronic complications; however, accuracy and stability issues are still under development, preventing a more widespread use in the market (Keneth, 2007).

3. Modeling Glucose Oxidase Biosensors

3.1. Glucose Oxidase Amperometric Biosensors

Nowadays there exist diverse techniques to measure glucose levels in the blood. The most common are spectrophotometric using small devices called *glucometers*. Some of these take advantage of the oxidation of glucose to gluconolactone catalyzed by glucose oxidase. Others use a similar reaction but with another enzyme, the glucose dehydrogenase. This latter brings up the advantage of more sensitivity but is less stable in presence of other substances. Some alternatives in glucose monitoring include control of the ketones in the urine, in case that glucose detection in blood becomes difficult (American Diabetes Association, 2013).

Currently, there exists an alternative monitoring technology based on electrochemical enzymatic sensors. They consist of the elements, biochemical and physical, assembled in direct contact, or close enough to establish a relationship with the analyte, to produce a measurable signal, as indicated in Figure 1. An enzymatic amperometric sensor works due to oxygen consumption, hydrogen peroxide production, or b-nicotinamide adenine obtaining during the process of the catalytic conversion of the substrate (Equations 1 and 2) (Prodromidis & Karayannis, 2002). The occurring electrochemical reactions are commented below.

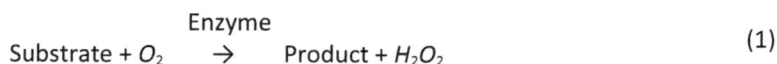

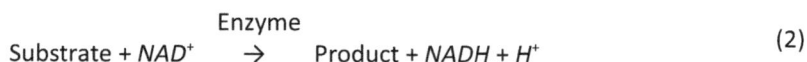

$$\text{Substrate} + O_2 \quad \xrightarrow{\text{Enzyme}} \quad \text{Product} + H_2O_2 \tag{1}$$

$$\text{Substrate} + NAD^+ \quad \xrightarrow{\text{Enzyme}} \quad \text{Product} + NADH + H^+ \tag{2}$$

The enzyme useful for measuring and assessing blood-glucose levels in biological fluids is known as *glucose oxidase*. The enzyme Glucose Oxidase belongs to the oxidoreductase class that catalyzes the β-D-glucose oxidation to D-glucone-1, 5-lactone and hydrogen peroxide (Equation 3). It is produced by the *Penicillium Notatum* and other fungi in the presence of glucose and oxygen. It is used to measure the glucose concentration in blood and urine samples. The general reaction equations can be described as:

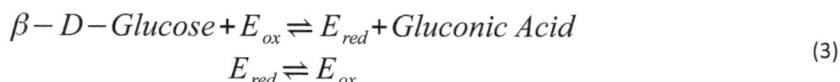

$$\beta - D - Glucose + E_{ox} \rightleftharpoons E_{red} + Gluconic\ Acid$$
$$E_{red} \rightleftharpoons E_{ox} \tag{3}$$

where E_{ox} and E_{red} are the oxidized and the reduced form of the glucose oxidase enzyme. Given this, there exist three generations of GOAB:

1st Generation. With O_2 added to the equation:

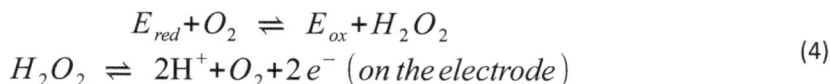

$$E_{red} + O_2 \rightleftharpoons E_{ox} + H_2O_2$$
$$H_2O_2 \rightleftharpoons 2H^+ + O_2 + 2e^- \ (on\ the\ electrode) \tag{4}$$

2nd Generation. A mediator is present:

$$E_{red} + M_{ox} \rightleftharpoons E_{ox} + M_{red}$$
$$M_{red} \rightleftharpoons M_{ox} + ne - (on\,the\,electrode)$$

(5)

3rd Generation. Without oxygen or other mediator:

$$E_{red} \rightleftharpoons E_{ox}$$

(6)

It is known that 1st Generation GOABs have a number of disadvantages, mainly oxygen concentration fluctuations; H_2O_2 inhibits the glucoseoxidase, among others. A solution that circumvents these problems was found in the 2nd and 3rd Generation GOABs (Stoytcheva, Nankov & Sharkova, 1995). This study focuses precisely in the 2nd generation GOABs.

3.2. GOAB Mathematical modeling

Mathematical modeling represents a powerful tool in the electrochemical biosensors design cycle, allowing for considerable reductions in costs and time (Wang, 2011). Despite the high benefits that GOAB technology brings to real life applications, it still bears some delicate design issues that make a better analysis and understanding still necessary (Petrauskas & Baronas, 2009). The GOAB response and its performance is affected by a number of factors, including electrode construction features (material, area and geometry), electrical factors (potential, current, charge, impedance), electrolytic factors (pH, solvents), and reaction variables (thermodynamic and kinetic parameters) (Borgmann, Schulte, Neugebauer & Schuhmann, 2011).

Given that Amperometric Electrochemical Biosensors (AEB) are used in combinatory synthesis procedures, the chemical reactions end-product might be scarce and limited, being measured from micrograms to miligrams scale; moreover, substrates inside enzymes can not be measured with analytical devices. Since the late 70s, several mathematical models have been used and proposed to this end, as an important tool looking for higher precision and simplicity (Malhotra & Chaubey, 2003). First AEB modeling efforts dealt with quantitative descriptions of biosensors kinetic behavior of simple idealized enzymes (Blaedel, Kissel & Boguslaski, 1972). Steady-state flux and distribution equations were used to show how enzymes fixed in gels may be used for immobilized enzyme kinetics analysis. The urease electrode potentiometric response were used in the experiment. Steady-state modeling of the current response by means of digital devices was one of the first digital simulation in glucose oxidase biosensor analysis (Mell & Maloy, 1975). Current modeling approaches range from analytical solutions of partial differential equations applied to simple biocatalytic processes to complex computer simulations of catalytic conversions and multiple transducer geometry (Baronas, 2010). Recent advances contemplate the use of *machine learning* (ML) algorithms, such as Artificial Neural Networks or Support Vector Machines. The use of these tools in AEB modeling is an emerging topic in the specialized scientific literature (Rangelova, Tsankova & Dimcheva, 2010; Alonso, Istamboulie, Ramírez-García Noguer, Marty & Muñoz, 2010). In this sense, model construction by ML techniques becomes a reasonable strategy, given that its black-box point of view liberates the modeler to fully and clearly express the mathematical laws underlying the physical phenomenon – in our case, the amperometric response of a biosensor.

4. Materials and Methods

4.1. Available data

Second Generation biosensors incorporate a mediator, in this case a p-benzoquinone mediated amperometric graphite sensor with covalently linked glucoseoxidase. This mediator is responsible for the electronic transfer between the enzyme and the electrode surface. Additionally, the following reagents were used: glucose oxidase (E.C. 1.1.3.4. from *Aspergillus*, 1000 U/mg), N-cyclohexyl-N'-[2-(methylmorpholino)ethyl]carbodiimide-4-toluenesulphonate (Merk) and glucose.

Amperometric data acquisition was achieved using a Radelkis OH-105 polarograph. The enzymatic working electrode used was a rotating disk electrode with a diameter of 6 mm, prepared from a spectrally pure graphite with glucoseoxidase immobilized on its surface. Saturated calomel electrode was used as reference electrode. The auxiliary electrode was a glassycarbon electrode.

The amperometric response was analyzed under different conditions of the Glucose *(Glucose)*, pH *(PH)*, temperature *(T)* and concentration of the mediator, the p-benzoquinone *(Benzoquinone)*.

Values for these input parameters (used as predictors) were Glucose (in mM) $\in$ {4, 8, 12, 16, 20}; p-benzoquinone (in mM) $\in$ {1, 0.8, 0.4, 0.2}; pH (dimensionless) $\in$ {4, 5, 6, 7} and Temperature (in Celsius scale) $\in$ {20, 37, 47, 57}. The response or faradaic current (I) was measured in mA.

The resulting data file consists of 320 rows (observations) and 5 columns (4 predictive variables and a continuous target variable, which corresponds to the biosensor response). As stated above, the predictive variables are: *Glucose, Benzoquinone, T* and *PH*. These predictive variables are standardized to zero mean, unit standard deviation. Finally the data file is shuffled to avoid predefined ordering biases.

4.2. Regression Methods

Suppose we are given training data of the form $\{(x_n, t_n)\}^N_{n=1} \subset X \times R$, where $X = R^d$ denotes the space of input vectors.

4.2.1. Classical regression analysis

Generalized linear models (GLMs) are a commonly used form of model for regression modelling. A GLM takes the form:

$$f_{GLM}(x) = \sum_{m=0}^{M} \beta\, m\Phi\, m(x) \qquad (7)$$

where φ is a set of M basis functions (which can be arbitrary real-valued functions) with $\varphi_0(\cdot) = 1$ and β is a vector of coefficients.

The number and form of the basis functions has to be decided beforehand. In this work we consider two useful GLM settings: linear and restricted polynomial regression:

Linear regression We consider the choices $\varphi_m(x) = x_m$ and $M = d$

Polynomial regression We consider the choices $\varphi_m(x)$ to be polynomials in x of limited degree (three, at most); in this case, every monomial will have its own β coefficient and M will be the total number of monomials.

The basis functions define a projection of the input data into a higher-dimensional space where the data is more likely to be linear. Since the obtained expressions are linear in the coefficients, these coefficients are optimized using standard least squares methods.

In order to assess how well these models fit the data, the leave-one-out cross-validation (LOOCV) method can be used. Every observation is excluded from the training set, the model is fit using the remaining points, and then is made to predict the left out observation. The process is repeated over the entire training set, and the LOOCV error is computed by taking the average over these predictions. This method provides an almost unbiased estimate of the generalization error and has the added advantage of being fast to compute for linear models. Typically one has to maximize:

$$R^2 cv = 1 - \frac{GE_{cv}(\hat{y})}{\frac{1}{N} \sum_{i=1}^{N} (\hat{y} - \overline{y})^2}$$

where $\quad GE_{cv}(\hat{y}) = \frac{1}{N} \sum_{i=1}^{N} \left(\frac{e_i}{1 - h_{ii}}\right)^2 \quad$ is the LOOCV error, and e_i, h_{ii} are the residual and the

leverage of observation x_i. A related measure of performance (to be minimized) is given by the normalized root mean-square error (NRMSE) (Bishop, 1996):

$$\sqrt{\frac{\sum_{i=1}^{N} \left(\frac{e_i}{1 - h_{ii}}\right)^2}{\sum_{i=1}^{N} (\hat{y} - \overline{y})^2}}$$

which can be interpreted as the fraction of output standard deviation that is explained by the model. Note that we can obtain the corresponding cross-validation NRMSE as $\quad \sqrt{1 - R^2_{cv}}$

4.2.2. Support Vector Machines

Support Vector Machines for regression (SVMR) have become a popular tool for the modelling of non-linear regression tasks (Smola, & Schölkopf, 2004). The SVMR is one of the several kernel-based techniques available in machine learning. These are methods based on implicitly mapping the data from the original input space to a feature space of higher dimensionality and then solving a linear problem in that space. The used E-insensitive loss function $|z|_\varepsilon = \max\{0, |z| - E\}$

penalizes errors that are greater than a threshold E, usually leading to sparser representations, entailing algorithmic and representational advantages, (Vapnik, 1998).

Let H be a real RKHS with kernel κ. The input data is transformed with a feature map $\Phi: X \rightarrow H$, to obtain the new data set $\{(\Phi(X_n), t_n)\}^N_{n=1}$. In a SVMR, the aim is to find a function $f_{SVMR}: (\Phi(x), w)_H + a_0$, for some $w \in H$ and $a_0 \in R$, which is as flat as possible and deviates a maximum of E from the given target values t_n, for all $n = 1,..., N$.

The usual formulation of the optimization problem is as the dual of the convex quadratic program:

$$\min_{w \in H, a_o \in R} \quad \frac{1}{2}\|w\|^2_h + \frac{c}{N}\sum_{n=1}^{N}\left(\xi_n + \hat{\xi}\right)$$

$$Subject\ to \begin{cases} \langle \Phi(x_n), w \rangle H + a_o - t_n \leq \varepsilon + \xi_n \\ t_n - \langle \Phi(x_n), w \rangle H - a_0 \leq \varepsilon + \xi_n \\ \xi_n, \hat{\xi}_n \geq 0 \end{cases} \tag{8}$$

for $n = 1, ..., N$. To solve (8), one considers the dual problem derived by the Lagrangian:

$$\max_{a, \hat{a}} \begin{vmatrix} -\frac{1}{2}\sum_{n,m=1}^{N}(\hat{a}_n - a_n)(\hat{a}_m - a_m)k(x_n, x_m) \\ -\epsilon\sum_{n=1}^{N}(\hat{a}_n - a_n) + \sum_{n=1}^{N}t_n(\hat{a}_n - a_n) \end{vmatrix} \tag{9}$$

$$Subject\ to \quad \sum_{n=1}^{N}(\hat{a}_n - a_n) = 0 \quad and \quad a_n, \hat{a}_n \in [0, C/N]$$

Exploiting the saddle point conditions, it can be proved that $\quad W = \sum_{n=1}^{N}(\hat{a}_n - a_n)\Phi(x_n)\quad$; given that $\quad k(x, x') = \langle \Phi(x), \Phi(x') \rangle H\quad$, the solution becomes'

$$fSVMR(x) = \sum_{n=1}^{N}(\hat{a}_n - a_n)k(x_n, x) + a_0, x \in X \tag{10}$$

4.2.3. Relevance Vector Machines

The Relevance Vector Machine (RVM) is a sparse Bayesian method for training GLMs which has the same functional form as the SVMR (Tipping, 2001). It is a kernel-based technique that typically leads to sparser models than the SVMR, and may also perform better in many cases. The RVM introduces a prior over the weights governed by a set of hyperparameters, one associated with each weight, whose most probable values are iteratively estimated from the data. The RVM has a reduced sensitivity to hyperparameter settings than the SVM.

In the RVM, a zero mean Gaussian prior with independent variances (acting as hyperparameters) $\alpha_j \equiv 1/\sigma^2_{w,j}$ is defined over each weight:

$$p(w|\alpha) = \prod_{j=1}^{M} \sqrt{\frac{\alpha_j}{2\pi}} \exp\left(\frac{-1}{2}\alpha_j w_j^2\right)$$

As in standard regression, assuming an independent zero-mean Gaussian noise model of variance σ^2 for the targets, the likelihood of a target vector t is:

$$p(t|w,\sigma^2) = (2\pi\sigma^2)^{\frac{-N}{2}} \exp\left(\frac{-1}{2\sigma^2}\|t - \Phi w\|^2\right)$$

where Φ is the Gram or kernel matrix of the inputs. In these conditions, the posterior over the weights $p(w|t,\alpha,\sigma^2)$ is also a Gaussian $N(\mu, \Sigma)$ that can be obtained using the Bayes rule:

$$p(w|t,\alpha,\sigma^2) = \int p(t|w,\sigma^2)\,p(w|\alpha)\,dw$$
$$p(w|t,\alpha,\sigma^2)$$

where $\mu = \sigma^{-2}\Sigma\Phi^T t$ and $\Sigma = (\sigma^{-2}\Phi^T\Phi + \Lambda)^{-1}$, being $\Lambda = \mathrm{diag}(\alpha_1,...,\alpha_M)$. Now the likelihood distribution over the training targets can be calculated by integrating out the weights to obtain the marginal likelihood for the hyperparameters:

$$p(t|\alpha,\sigma^2) = \int p(t|w,\sigma^2)p(w|\alpha)d\,w$$

This marginal distribution is again Gaussian $N(\mathbf{0}, A)$, where $A = \sigma^2 I + \Phi\Lambda^{-1}\Phi^T$. For computational efficiency, the logarithm of the evidence is maximized:

$$L(\alpha) = \log p(t|\alpha,\sigma^2) = -\frac{1}{2}\left(M\log(2\pi) + \log|A| + t^T A^{-1} t\right)$$

The estimated value of the model weights is given by their maximum a posteriori (MAP) estimate, which is the mean of the posterior distribution $p(w|t,\alpha,\sigma^2)$. This MAP estimate depends on the hyperparameters α and σ^2. These variables are obtained by maximizing the marginal likelihood $L(\alpha)$. Sparsity is achieved because in practice many of the hyperparameters α_j tend to infinity, yielding a posterior distribution of the corresponding weight w_j that is sharply peaked around zero. These weights can then be deleted from the model, as well as their associated basis functions.

5. Experimental setup

The experimental part explores the modeling of the biosensor output from two different points of view. In the first set of experiments, we treat the inputs as they are, namely continuous regressors. In the second set of experiments, we consider the possibility of treating the regressors as categorical instead of continuous. This is supported by the fact that all four predictors take on a limited number of values (to be precise, four for the *Benzoquinone*, *T* and *PH*, and five for the *Glucose*). Categorical predictor variables cannot be entered directly into a regression model (and be meaningfully interpreted). Typically, a categorical variable with *c* modalities will be transformed into *c* – 1 binary variables (each with two modalities).

For example, if a categorical variable had five modalities, then four binary variables are created that would contain the same information as the single categorical variable. In particular, all the distances between the modalities are equal, regardless of the specific coding chosen. These variables have the advantage of simplicity of interpretation and are sometimes preferred to correlated predictor variables. They are also useful to assess non-linearities between the regressors and the output.[*] In the present situation, the new values for the *Glucose* are "very low", "low", "medium", "high" and "very high" (64 observations each), whereas new values for the other three predictors are "low", "low-medium", "medium-high" and "high" (80 observations each).

5.1. Optimization of the SVMR and the RVM

First, the available data were randomly split into two sets: 220 observations (68.75%) for training and the remaining 100 observations (31.25%) for testing.

The standard regression methods need no additional parameter specification. In order to obtain the solution for a kernel method, one has to choose the kernel function, and determine appropriate values for the associated hyperparameters. In the *SVMR*, the *E* parameter controls the width of the *E*-insensitive zone. The cost parameter *C* determines the trade-off between model complexity (flatness of the solution) and the degree to which deviations larger than E are tolerated. The value of *E* can affect the number of support vectors used to construct the regression function. The bigger the *E*, the fewer support vectors are selected and smoother regression functions. An effective approach is to estimate the generalization error –usually through cross-validation– and then optimize for these parameters so that this estimation is minimized. For the *SVMR*, *C* is varied logarithmically between $10^{-1.5}$ and $10^{1.5}$ (20 equally-spaced values for the exponent), and E is varied logarithmically between 10^{-3} and 10^{0} (10 equally-spaced values for the exponent).

Among the kernels that are available in the literature, we select the polynomial kernel: the function

$$k_{Poly}(x,y) = (\langle x,y \rangle + R)^{m}$$

is called the polynomial kernel, with $R \geq 0$ and integral $m \geq 1$.

---

[*] In rigour these are categorical *ordinal* predictors, although we will treat them as categorical *nominal* ones, given the absence of specific methods for ordinal predictors (Agresti, 2002).

Another kernel that can be used is the Gaussian Radial Basis Function (RBF) kernel, known to be a safe default choice for kernel methods working on real vectors (Schölkopf & Smola, 2001):

$$k_{RBF}(x, y) = \exp \quad (-\gamma \|x - y\|^2)$$ (11)

where $\gamma > 0$ is the smoothing parameter. The γ parameter in the *RBF* kernel is estimated using the *sigest* method, based upon the 10% and 90% quantiles of the sample distriubion of $\|x - x'\|^2$ (Caputo, Sim, Furesjo & Smola, 2002). We try different polynomial kernels, given by degrees *m* from 1 to 5 and $R = 1$.

The parameters were optimized through 30 times 10-fold cross-validation (30 x 10 cv) usingthe training set; a model is then refit in the training set using the best parameter configuration; this model is now made to predict the held out test set. The error reported in all cases is the normalized root mean-square error (NRMSE) in the test set. A model showing a NRMSE of 1 corresponds to the best *constant* regression; good models should then have a NRMSE considerably smaller than 1 (and reaso`ably close to 0).

For the RVM, the same consideratiors apply for the kernels and their parameters (the RVM needs no specification of *C* or the *E* parameter). Theoretically, the whole training set could be used to fit the RVM (without cross-valication). However, resampling is still needed to choose the best kernel configuration; therefore, the same 30 x 10 cv procedure is used to evaluate performance in the training set.

6. Results and Discussion

6.1. Basic statistical analysis

After the pre-process described in section 4.1, we get a data set with the summary described in Table 1. The 'Target' variable refers to the biosensor output.

Glucose	*Benzoquinone*	*T*	*PH*	*Target*
Min.: -1.412	Min.: -1.2629	Min.: -1.4797	Min.: -1.3395	Min.: 0.2848
1st Qu.: -0.706	1st Qu.: -0.7893	1st Qu.: -0.5481	1st Qu.: -0.6698	1st Qu.: 1.9575
Median: 0.000	Median: 0.0000	Median: 0.1279	Median: 0.0000	Median: 4.2178
Mean: 0.000	Mean: 0.0000	Mean: 0.0000	Mean: 0.0000	Mean: 12.3994
3rd Qu.: 0.706	3rd Qu.: 0.7893	3rd Qu.: 0.6759	3rd Qu.: 0.6698	3rd Qu.: 14.9122
Max.: 1.412	Max.: 1.2629	Max.: 1.2240	Max.: 1.3395	Max.: 75.5506

Table 1. Descriptive statistics after pre-processing

We can see that all the predictive variables are perfectly symmetrical (the mean and median are equal), with the exception of *T*, whose distribution is skewed to the left (negative skew), since its mean is smaller than its median (see the set of boxplots in Figure 2).

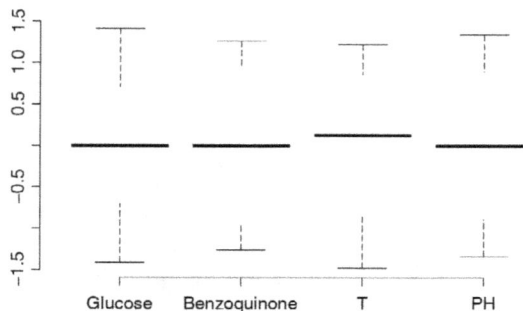

Figure 2. Box plots of predictive variables

A convenient first step is to take the natural log of the target variable. The effect of this change is clearly visible (Figure 3).

Now we compute the Pearson's product-moment correlations (Table 2).

First variable	Second variable	Correlation
PH	Target	-0.332
Glucose	Target	0.267
T	Target	0.167
Benzoquinone	Target	0.063

Table 2. Pearson's variable correlations; only the four largest are shown

These very small correlations suggest that only *PH* bears some linear relation with the target variable,[*] in addition, there is no linear relation concerning the predictive variables with one another. In order to refine this result, we compute Spearman's ρ correlation coefficient. Although this coefficient does not detect general quadratic non-linearities, it is good for detecting possible outliers and monotonic non-linearities (Table 3).

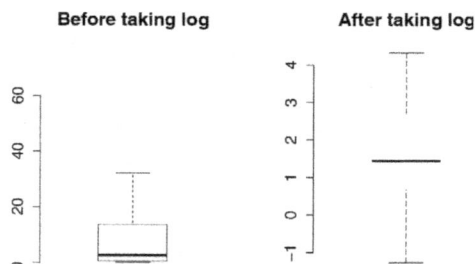

Figure 3.Box plots of the target (output) variable

---

[*]   Although true correlations are not equal to 0 (*PH*-Target: t = -6.2685, df = 318, p-value = 1.187e-09, *Glucose*-Target: t = 4.9337, df = 318, p-value = 1.303e-06, *T*-Target: t = 3.0135, df = 318, p-value = 0.00279).

First variable	Second variable	Correlation
PH	Target	-0.569
Glucose	Target	0.405
T	Target	0.195
Benzoquinone	Target	0.082

Table 3. Spearman's variable correlations; only the four largest are shown

The correlations are larger for some variables –notably *PH* and the *Glucose*, suggesting a relation of non-linear nature with the biosensor output.

6.2. Regression with continuous predictors

6.2.1. Standard Linear Regression

	Estimate	Std. Error	*t* value	Pr(>|t|)
(Intercept)	1.6182	0.0653	24.7764	6.3e-65
Glucose	0.5072	0.0666	7.6171	8.1e-13
Benzoquinone	0.1151	0.0658	1.7480	8.2e-02
T	0.3436	0.0650	5.2883	3.0e-07
PH	-0.6896	0.0647	-10.6570	1.5e-21

Table 4. Coefficients and significance of the linear regression analysis with continuous predictors

The results of linear regression are shown in Table 4. These results show that all the coefficients are significantly different from zero, as given by the negligible p-values, with the exception of that for *Benzoquinone*, which is barely significant (p-value = 0.082). Since the predictive variables are standardized, the coefficients can be related to the relevance of the corresponding variables. The most important predictor is *PH*, followed by *Glucose* and *T*; the *Benzoquinone* is by far the less important predictor. The importance of variables can be further assessed by an Analysis of Variance (ANOVA) (Table 5).

	Df	Sum Sq	Mean Sq	*F* Value	Pr(>F)
Glucose	1	54.6828	54.6828	58.5062	6.7e-13
Benzoquinone	1	1.0242	1.0242	1.0958	3.0e-01
T	1	24.7466	24.7466	26.4768	6.0e-07
PH	1	106.1500	106.1500	113.5718	1.5e-21
Residuals	215	200.9500	0.9347		

Table 5. ANOVA in the training set with continuous predictors

According to the ANOVA, PH is by large the most important predictor, followed by *Glucose* and *T;* the *Benzoquinone* is by far the less important predictor. However, an AIC analysis does not suggest to remove any of the regressors, and therefore we keep all of them. We may now wish to see how well the model fits and predicts the training data. We get a relatively low R^2_{cv} = 0.4724 (corresponding to a NRMSE of 0.7264), which indicates a rather poor model.

The previous results using linear regression again suggest that the possible relation between the output of the biosensor and the predictive variables is a non-linear one. Therefore polynomial regression as described in section 4.2 is considered. After preliminary modeling trials in the training set, one ends up with a set of regressors formed by a third-degree polynomial on *PH* and a second-degree polynomial on both *T* and the *Glucose*. The addition of other terms does not increase the model quality and adds further complexity. The results of this polynomial regression are shown in Table 6.

	Estimate	Std. Error	*t* value	Pr(>|*t*|)
(Intercept)	1.6468	0.0103	159.1228	1.0e-221
poly(*PH*, 3)-1	-12.9545	0.1831	-70.7585	5.8e-149
poly(*PH*, 3)-2	-7.6297	0.1857	-41.0931	1.2e-102
poly(*PH*, 3)-3	13.3168	0.1871	71.1572	1.9e-149
poly(*T*, 2)-1	5.7589	0.1839	31.3108	3.0e-81
poly(*T*, 2)-2	-5.1793	0.1836	-28.2150	1.5e-73
Benzoquinone	0.1012	0.0105	9.6714	1.5e-18
poly(*Glucose*, 2)-1	9.3897	0.1882	49.8934	9.0e-119
poly(*Glucose*, 2)-2	-3.5097	0.1842	-19.0517	9.7e-48

Table 6. Coefficients of the non-linear (polynomial) regression analysis with continuous predictors. The notation poly(V, r)-s stands for the s-degree monomial of an r-degree polynomial on regressor V

The results show that all the coefficients are significantly different from zero, as given by the negligible *p*-values, with no exception. Indeed, the AIC analysis does not suggest to remove any of the regressors. Variable importance can again be further assessed by the ANOVA (Table 7).

We may now wish to see how well the model fits and predicts the training data. We get an excellent R^2_{cv} = 0.9872 (corresponding to a NRMSE of only 0.1132), which indicates a very promising model, remarkably better than the linear one.

	Df	Sum Sq	Mean Sq	F Value	Pr(>F)
poly(*PH*, 3)	3	273.3378	91.1126	3899.0107	1.9e-184
poly(*T*, 2)	2	40.3336	20.1668	863.0042	2.6e-102
Benzoquinone	1	2.3375	2.3375	100.0308	1.6e-19
poly(*Glucose*, 2)	2	66.6140	33.3070	1425.3161	2.8e-123
Residuals	211	4.9307	0.0234		

Table 7. ANOVA in the training set with continuous predictors. The notation poly(V, r) stands for an r-degree polynomial on regressor V

6.2.2. Regression with the SVMR

We now turn to results obtained with the kernel-based methods (Table 8). Using a linear kernel, the best set of parameters for the SVMR (optimized through 30 x 10 cv) is *E* = 0.464 and *C* = 0.695. The cross-validation NRMSE of this choice is 0.7387. Actually all the results are in the range 0.73-0.80, no matter the value of *C* and *E*. Something similar happens for the quadratic kernel, although in this case the range is 0.60-0.80. These readings suggest a general poor model, in consonance with the linear regression model previously reported. Using higher degree kernels, the results are much better, in accordance to those found for the polynomial regression model. Indeed, the result with a cubic polynomial is similar to the polynomial regression (which used cubic polynomials in the *PH*). The best result corresponds to the quartic polynomial, markedly better than the polynomial regression.

Kernel	degree	*C*	*E*	cv NRMSE
Linear	1	0.695	0.464	0.7387
Quadratic	2	0.036	0.464	0.6017
Cub c	3	1.623	0.046	0.0867
Quartic	4	0.215	0.046	0.0491
Quintic	5	0.183	0.046	0.0942
RBF	-	0.069	0.215	0.4441

Table 8. Results for the SVMR with polynomial kernels of different degrees and the RBF kernel with continuous predictors; cv NRMSE is the cross-validation NRMSE

The results seem to confirm the need for at least third-order information extracted from the original regressors; however, higher-order terms start to overfit the data. The relatively poor results obtained by the RBF kernel can also be explained in this light. A nice aspect of the results is that, for all kernels, for same values of *E*, predictive performance is tied for many values of the *C* parameter; in these cases, we selected the smallest value of *C*, in accordance with Statistical Learning Theory (Vapnik, 1998).

6.2.3. Regression with the RVM

The results for the RVM with the different kernels are displayed in Table 9. They are in consonance with those for the SVMR previously reported, although they are consistently better. Remarkably, there is a coincidence with the SVMR in that the best result corresponds to the quartic polynomial.

Kernel	degree	cv NRMSE
Linear	1	0.7356
Quadratic	2	0.5973
Cubic	3	0.0845
Quartic	4	0.0301
Quintic	5	0.0392
RBF	-	0.2148

Table 9. Results for the RVM with polynomial kernels of different degrees and the RBF kernel with continuous predictors; cv NRMSE is the cross-validation NRMSE

6.3. Regression with categorical predictors

6.3.1. Standard Linear Regression

The results of linear regression are shown in Table 10.

	Estimate	Std. Error	*t* value	Pr(>|*t*|)
(Intercept)	1.1474	0.0364	31.5440	8.3e-81
Glucose (low)	-0.6432	0.0295	-21.8069	2.0e-55
Glucose (medium)	-0.1810	0.0286	-6.3219	1.6e-09
Glucose (very high)	0.1067	0.0298	3.5778	4.3e-04
Glucose (very low)	-1.4215	0.0298	-47.7724	2.1e-113
Benzoquinone (low)	-0.2697	0.0268	-10.0788	1.1e-19
Benzoquinone (low-medium)	-0.0804	0.0260	-3.0866	2.3e-03
Benzoquinone (medium-high)	0.0220	0.0267	0.8220	4.1e-01
T (low)	-0.7716	0.0260	-29.6715	2.6e-76
T (low-medium)	0.1902	0.0257	7.4008	3.4e-12
T (medium-high)	0.3650	0.0271	13.4747	4.3e-30
PH (low)	1.2709	0.0258	49.2501	6.4e-116
PH (low-medium)	2.8093	0.0268	104.6867	1.4e-180
PH (medium-high)	0.1746	0.0265	6.5778	3.9e-10

Table 10. Coefficients of the linear regression analysis with categorical predictors in the training set

These results show that all the coefficients are significantly different from zero, as given by the negligible p-values, with the exception of that for *Benzoquinone* (medium-high), which is not significant. The importance of variables can be further assessed by the ANOVA on the previous regression (Table 11).

According to the ANOVA, PH is by large the most important predictor, followed by Glucose and T; the *Benzoquinone* is by far the less important predictor. Again, AIC analysis does not suggest to remove any of the regressors, and therefore we keep all of them. We may now wish to see how well the model fits and predicts the training data. We get an excellent R^2_{cv} = 0.9887 (corresponding to a NRMSE of 0.1062). This result vastly improves that of linear regression with continuous predictors (NRMSE of 0.7264). Sadly, there is no possibility of developing a standard polynomial regression model using categorical predictors.

	Df	Sum Sq	Mean Sq	F value	Pr(>F)
Glucose	4	66.2744	16.5686	891.6448	8.5e-129
Benzoquinone	3	4.1249	1.3750	73.9946	1.6e-32
T	3	52.4748	17.4916	941.3168	6.1e-120
PH	3	260.8515	86.9505	4679.2708	3.6e-189
Residuals	206	3.8279	0.0186		

Table 11. ANOVA in the training set with categorical predictors

6.3.2. Regression with the SVMR

We now turn to results obtained with the kernel-based methods (Table 12). The best set of parameters is $E = 10^{-3}$ and C around 1, using the quadratic kernel. Actually all the results are markedly better than those obtained with the SVMR in the same conditions but using continuous predictors. This result is in consorance with that previously reported for the standard regression. Again, using non-linear (higher degree) kernels, the results are better than with linear ones, although in this case the technique starts to overfit at cubic polynomials. The RBF kernel performs much better too, and is comparable to the cubic polynomial.

Kernel	degree	C	E	cv NRMSE
Linear	1	3.793	0.1	0.1077
Quadratic	2	1.062	0.001	0.0026
Cubic	3	0.455	0.001	0.0307
Quartic	4	0.195	0.0022	0.1065
Quintic	5	0.127	0.001	0.2000
RBF	-	5.796	0.001	0.0371

Table 12. Results for the SVMR with polynomial kernels of different degrees and the RBF kernel with categorical predictors; cv NRMSE is the cross-validation NRMSE

The results make perfect sense in the light of model complexity, the linear kernel being too simple, and polynomials beyond the cubic one being too complex. In addition, it was observed that E was the critical parameter; for all kernels, for similar values of E, predictive performance varies smoothly with the C parameter, and many times it is rather independent; again, in these cases, we selected the smallest value of C.

6.3.3. Regression with the RVM

The results for the RVM with the different kernels are displayed in Table 13. They are in consonance with those for the SVMR previously reported, although this time those for the SVMR seem slightly better. Again, there is a coincidence with the SVMR in that the best result corresponds to the quartic polynomial.

6.4. Discussion

In view of the results reported so far, two methods stand out from the rest: the two nonlinear kernel methods of moderate complexity. Specifically, both the SVMR and the RVM with quadratic kernel and using categorical predictors deliver very good 30 x 10 cv errors, around or below 10^{-3} of NRMSE. The decision among these two methods is not an easy one, given that the errors are similar (with a slight advantage of the SVMR). On the other hand, the RVM is expected to deliver a sparser model. In order to decide, we proceeded to refit both methods in the training set using the best parameter configuration. The SVMR ($C = 1.062$ and $E = 10^{-3}$) delivers a modeling NRMSE of 0.0023 with 142 support vectors (a 64.5% of the training data); the RVM delivers a modeling NRMSE of 0.0020 with 63 relevance vectors (a 28.6% of the training data). Now the decision is clear: given that the models now have the right complexity (because they have been optimized towards minimizing predictive error), the modeling error is a relevant quantity. The RVM then achieves a smaller NRMSE with less than half of the regressors than the SVMR does.

Kernel	degree	cv NRMSE
Linear	1	0.1066
Quadratic	2	0.0030
Cubic	3	0.0371
Quartic	4	0.1133
Quintic	5	0.2239
RBF	-	0.0454

Table 13. Results for the RVM with polynomial kernels of different degrees and the RBF kernel with categorical predictors; cv NRMSE is the cross-validation NRMSE

Therefore we make the choice of the RVM with a quadratic kernel and categorical predictors. This model is then made to predict the held out test set, yielding a predictive NRMSE of 0.0022. This is a very nice result for two reasons:

- The predictive NRMSE of 0.0022 is equivalent to a residual (non-explained) variance of only 0.22% of the total variance of the (test) predicted target values; therefore the model is indeed a very accurate one.

- The training (or modeling) NRMSE of 0.0020 and the predictive NRMSE of 0.0022 are in very good agreement, and an indication of a model of the right complexity.

The predictive results can also be displayed. In Figure 4 the predictions are plotted against the true values. It can be seen that the predictions are very good even when expressed in the original units (exponentiating the prediction). To be precise, the prediction error expressed in the original units amounts to an NRMSE of 0.0027. In order to obtain a final model that could be used in the future, we refit the RVM with a quadratic kernel (using categorical predictors) in the entire dataset. The obtained model has a modeling NRMSE of 0.00184 with 62 relevance vectors (a 19.4% of the training data). Expressed in the original units, this error corresponds to a NRMSE of 0.00176.

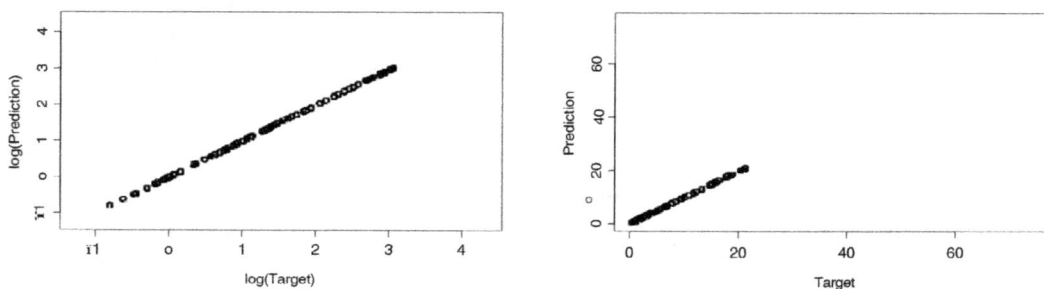

Figure 4. Final test set predictions. Left: in the log units; right: in the original units

At this point it is of great interest to make a comparison of all the results for the different regression methods and types of modeling.

Linear vs. non-linear methods. Given the poor (and consistent) results obtained by the three linear methods (linear regression, linear SVMR and linear RVM), it can be concluded that the true relation between the predictive variables and the biosensor output is a non-linear one.

Classical regression vs. kernel methods. Using continuous predictors, polynomial regression is able to give a fairly good result (NRMSE of 0.1132); kernel regression with the SVMR is able to improve this to half the error (NRMSE of 0.0491); kernel regression with the RVM delivers an NRMSE of 0.0301, both using a fourth-degree polynomial.

Continuous vs. categorical predictors. Although the non-linear methods make good use of the continuous variables, the limited amount of information they give (only 4 to 5 different values) makes the modeling a non-trivial undertaking; in contrast, standard regression and both the SVMR and the RVM deliver consistently better results using the categorized version of the dataset.

SVMR vs. the RVM Both methods can be seen as sparse GLM trainers, and indeed the RVM was initially presented as an alternative (and direct competitor) to the SVMR. The results follow similar paths for same kernels and it is not clear which one is performing better; the SVMR shows slightly lower errors but the RVM needs no parameter specification and delivers a sparser solution.

7. Conclusions

The continuous amperometric response of a GOAB has been successfully modelled by means of several classic and statistical learning methods. Specifically, kernel-based regression techniques have been used. The reported experimental results show a very low prediction error of the biosensor output, obtained using a relevance vector machine (RVM) with a quadratic kernel and categorical predictors. This constitutes a rather simple model of the biosensor output, because it is sparse and uses only four regressors with a limited number of values. We have also found that the pH is the most important predictor, followed by the Glucose and the temperature. An alternative technique could be grounded in the support vector machine for regression (SVMR) While SVMR can only be applied to the subset of generalized linear models (GLM) that can be defined by a valid kernel function, the RVM can train a GLM with any collection of basis functions.

Continuous glucose monitoring by means of a GOAB can constitute a remarkable ally to diabetic patients involved in serious collateral chronic complications. However, their design is still under development, in order to improve both accuracy and stability. In electrochemical biosensors design, mathematical modeling is a highly recurrent tool, given that it facilitates the computational simulation saving design and testing time and resources. The experimental proposal and conditions developed in this chapter could be applied for other scenarios in the wide spectrum of biosensing technology.

References

Agresti, A. (2002). *Categorical Data Analysis*. Wiley Series in Probability and Statistics, 2nd edition. Wiley-Interscience.

Alonso, G., Istamboulie, G., Ramírez-García A., Noguer, T., Marty, J., & Muñoz, J. (2010). Artificial neural network implementation in single low-cost chip for the detection of insecticides by modeling of screen-printed enzymatic sensors response. *Computers and Electronics in Agriculture, 74(2)*, 223-229. http://dx.doi.org/10.1016/j.compag.2010.08.003

American Diabetes Association. (2013). Available at http://www.diabetes.org

Baronas, R. (2010). *Mathematical modeling of biosensors an introduction for chemists and mathematicians*. Springer. http://dx.doi.org/10.1007/978-90-481-3243-0

Bishop, C. (1996). *Neural Networks for Pattern Recognition*. Oxford University Press, USA.

Blaedel, W.J., Kissel, T.R., & Boguslaski, R.C. (1972). Kinetic behavior of enzymes immobilized in artificial membranes. *Analytical Chemistry, 44(12)*, 2030-2037. PMID: 4657296. http://dx.doi.org/10.1021/ac60320a021

Borgmann, S., Schulte, A., Neugebauer, S., & Schuhmann, W. (2011). Amperometric biosensors. In R. Alkire, D. Kolb & J. Lipkowski (Eds.). *Bioelectrochemistry: Fundamentals, Applications and Recent Developments*. Wiley-VCH.

Caputo, B., Sim, K., Furesjo, F., & Smola, A. (2002). Appearance-based Object Recognition using SVMs: Which Kernel Should I Use. *Proc. of NIPS workshop on Statistical methods for computational experiments in visual processing and computer vision*.

Department of Trade and industry. (2012). *Biosensors for Industrial Applications, A review of Biosensor Technology*. United Kingdom Goverment.

Diabetes Research Institute. (2012). Available at http://www.diabetesresearch.org

Elliott, C. (2006). Biosensor detects toxic drugs in food. *Trac-trends in Analytical Chemistry, 25*.

Keneth, W. (2007). How to design a biosensor. *Journal of Diabetes Science and Technology, 1(2)*, 201-204. http://dx.doi.org/10.1177/193229680700100210

Malhotra, B., & Chaubey, A. (2003). Biosensors for clinical diagnostics industry. *Sensors and Actuators B: Chemical, 91(1-3)*, 17-127. http://dx.doi.org/10.1016/S0925-4005(03)00075-3

Mell, L.D., & Maloy, J.T. (1975). Model for the amperometric enzyme electrode obtained through digital simulation and applied to the immobilized glucose oxidase system. *Analytical Chemistry, 47(2)*, 299-307. http://dx.doi.org/10.1021/ac60352a006

Morrison, D., Dokmeci, M., Demirci, U., & Khademhosseini, A. (2008). Clinical applications of micro- and nanoscale biosensors. In K. Gonsalves, C. Halberstadt, C. Laurencin & L. Nair (Eds.). *Biomedical Nanostructures*. Wiley.

Petrauskas, K., & Baronas, R. (2009). Computational modelling of biosensors with an outer perforated membrane. *Nonlinear Analysis: Modelling and Control, 14(1)*, 85-102.

Pohanka, M., Skládal, P., & Kroca, M. (2007). Biosensors for biological warfare agent detection. *Defence Science Journal, 57.*

Prodromidis, M., & Karayannis, M. (2002). Enzyme based amperometric biosensors for food analysis. *Electroanalysis, 14(4)*, 241-261. http://dx.doi.org/10.1002/1521-4109(200202)14:4<241::AID-ELAN241>3.0.CO;2-P

Rangelova, V., Tsankova, D., & Dimcheva, N. (2010). Soft computing techniques in modelling the influence of ph and temperature on dopamine biosensor. In V. Somerset (Ed.). *Intelligent and Biosensors*. INTECH. http://dx.doi.org/10.5772/7029

Saharudin, H., & Asim, R. (2006). Optical biodetection of cadmium and lead ions in water. *Medical engineering & physics, 28(10)*, 978-981. http://dx.doi.org/10.1016/j.medengphy.2006.04.004

Schölkopf, B., & Smola, A. (2001). *Learning with Kernels: Support Vector Machines, Regularization, Optimization, and Beyond*. Cambridge, MA, USA: MIT Press.

Smola, A., & Schölkopf, B. (2004). A tutorial on support vector regression. *Statistics and Computing, 14(3)*, 199-222. http://dx.doi.org/10.1023/B:STCO.0000035301.49549.88

Stoytcheva, M., Nankov, N., & Sharkova, V. (1995). Analytical characterisation and application of a p-benzoquinone mediated amperometric graphite sensor with covalently linked glucoseoxidase. *Analytica Chimica Acta, 315(12)*, 101-107. http://dx.doi.org/10.1016/0003-2670(95)00314-P

Thévenot, D., Toth, K., Durst, R., & Wilson, G. (2001). Electrochemical biosensors: recommended definitions and classification. *Biosensors and Bioelectronics, 16(1-2)*, 121-131. http://dx.doi.org/10.1016/S0956-5663(01)00115-4

Tipping, M. (2001). Sparse bayesian learning and the relevance vector machine. *Journal of Machine Learning Research, 1*, 211-244.

Vapnik, V. (1998). *Statistical learning theory*. Wiley.

Wang, Q (2011). *Mathematical Methods for Biosensor models*. PhD thesis, Dublin Institute of Technology.

World Health Organization. (2013). Available at http://www.who.int

Chapter 9

Modelling the influence of pH and temperature on the response of an acetylcholinesterase biosensor using machine learning methods

Edwin R. García Curiel, Larysa Burtseva, Margarita Stilianova Stoytcheva, Félix F. González-Navarro, Ana Sofia Estrella Sato

Universidad Autónoma de Baja California, México.

edwin.garcia@uabc.edu.mx, burtseva@uabc.edu.mx, margarita.stoytcheva@uabc.edu.mx, fernando.gonzalez@uabc.edu.mx, sofia.estrella@uabc.edu.mx

Dci: http://dx.doi.org/10.3926/oms.189

Referencing this chapter

García Curiel, E.R, Burtseva, L., Stilianova-Stoytcheva, M., González-Navarro, F.F., & Estrella Sato, A.S. (2014). Modelling the influence of pH and Temperature on the response of an acetylcholinesterase biosensor using Machine Learning Methods. In M. Stoytcheva & J.F. Osma (Eds.). *Biosensors: Recent Advances and Mathematical Challenges*. Barcelona: España, OmniaScience. pp. 185-202.

1. Introduction

Among the chemicals that damage the environment, the ones based on the inhibitions of the enzyme acetylcholinesterase (AChE) are the predominant agriculture insecticides, carbamates or organophosphates, but they bring serious health and environment risks. The use of electrochemical biosensors is the most commonly way of detection for AChE inhibitors based on carbamates. A biosensor is a device capable to produce an electrical signal resulting from a chemical reaction between a biological compound and any other substance, producing valuable information that can be analyzed.

The induced chemical reaction (between the enzyme AChE and the neurotransmitter acetylcholine (ACh) is highly dependent of external factors, such as pH and temperature; they affect the reaction and similarly, affect the biosensor readings –i.e its accuracy and the observed electrical power. The pH takes different values, one of which shows the largest electric power and is considered as the optimum value. In the case, of temperature, the electric current has an exponential behavior, however it reaches a maximum point around 60°C, where the enzyme undergoes a denaturation process causing the current decrease until disappears.

Since the current produced by the chemical reaction depends on the interaction of many variables, it cannot be explained by simple observation; therefore a computational approach is needed in order to model the faradaic current behavior. So far, several efforts have been made to model the resulting current using different biosensors:

- Digital simulation of the current response in steady state obtained by an amperometric biosensor for a glucose system; it maintains a solid mathematical basis, but the development of the model is committed to low electrical simulations, otherwise the margin of error increases (Mell & Malloy, 1975).

- Modeling of the resulting current from an electrochemical oxygen biosensor, however it only shows the models used and do not describe how it was implemented or tested (Rangelova, Tsankova & Dimcheva, 2010).

- A chip implementation using a neural network to simulate the current on an enzymatic-biosensor; it shows the process used, but not the number of samples used to test the model, or how the experiments were performed (Alonso, Istamboulie, Ramirez-Garcia, Noguer, Marty & Muñoz, 2010).

In this work, we step aside from these traditional approaches by tanking advantage of powerful computational regression models, whose theoretical background comes from the statistical pattern recognition field.

The proposal explained trough this contribution embodies several stages briefed as follows:

- Data analysis
- Pre-processing
- Algorithm settings
- Training
- Validation

As well as its evaluation:

- Tests

- Comparative analysis

Given an electric charge produced by the interaction between the enzyme AChE and the substrate Ach, with specific temperature and pH; the regression model is selected by a comparative analysis of different regression algorithms trough a certain performance measure explained later. Under this general approach, the response of an acetylcholinesterase biosensor will be studied, and a final mathematical model will be explained.

2. The Biosensors

2.1. Electrochemical Biosensor

According to the IUPAC (International Union of Pure and Applied Chemistry) definition, a biosensor is an analytical device that combines a biological component for molecular recognition and a signal processing device called a transducer, which can detect and measure, quickly and accurately, the signal produced by the interaction of the biological element and the substance of interest. The transducer, which normally ensures high efficiency of the sensor, can be thermal, optical, magnetic, nano-mechanic, piezoelectric or electrochemical. Furthermore, the selectivity of detection is ensured by a biological recognition element, which is based on a bio-ligand deoxyribonucleic acid (DNA, RNA ribonucleic acid, antibodies, etc.), or a biocatalyst (some redox proteins, individual enzymes and enzymatic systems, such as cell membranes, complete microorganisms) (Thevenot, Tóth, Durst & Wilson, 1999). Table 1 shows several classification criteria for biosensors depending on:

- Interaction type between the components to be detected;

- Interaction detecting method;

- Biological element to recognize;

- Device transducer type.

Interaction Type	Interaction Detecting Method
Biocatalitic	Direct
Bioafinity	Indirect
Biological Element to Recognize	**Transducer Type**
Enzyme	Electrochemical
Organelle, tissue or complete cell	Optic
Biological receptor	Piezoelectric
Antibody	Thermometric
Nucleic acid	Nanomechanic
Aptamers	

Table 1. Biosensors classification criteria

The electrochemical biosensors are divided into two types:

- Potentiometric.

- Amperometric.

Biosensors that measure only a change in potential at the interface-analyte sensor with respect to the reference electrode are known as potentiometric sensors or biosensors. Sensors that impose external potential to effect the transformation are referred to as electrochemical *amperometric biosensors*. Inhibition of the enzyme in these biosensors is monitored by the current change, detecting to a certain potential oxidation or reduction. Table 2 describes various environmental biosensors according to the type of transducer used on certain biological elements, compound to detect and the area possibly analyzed.

Transducer type	Biological element	Analite	Area
Electrochemical	Antibodies	Atrazine	*
Optic	Antibodies	Simazine	Ground, water mass
Optic	Antibodies	Pesticides	Rivers
Electrochemical	Antibodies	Tensoactives (alkylfenols)	*
Electrochemical	Antibodies	Estradiol	*
Electrochemical	Antibodies	Eschenchacoll	Drinking water
Optic	Antibodies	Enteritis Listeria monocytogenes Salmonella	*
Acoustic	Antibodies	Salmonella typhymurium	*
Optic	Enzyme (AChE)	Organophosphates compounds	Water
Electrochemical	Enzyme (AChE)	Paraoxon y carbofuran (pesticides)	Residual waters
Electrochemical	Enzyme (tyrosinase)	Fenols	Ground, dirt, waters

Table 2. Biosensors with environmental applications

2.2. Enzyme Reaction

The measurements were performed using an electrochemical biosensor with two electrodes, a counter electrode and a reference electrode in conjunction with an insulated electrode (working electrode), which in its lower part, has a portion (mg/cm^3) of the enzyme AChE. Figure 1 shows the schematic AChE biosensor used in the experiment.

The biosensor (working electrode) used in the experiment was prepared by immobilizing the AChE enzyme, chemically bonded on the surface of a graphite electrode. The auxiliary electrode was glassy-carbon type and the reference electrode a saturated calomel.

Thereafter the electrodes are immersed into a conventional electrochemical cell, which contains a buffer solution with a given concentration ($mmols/L$) of the substrate ACh. The solution receives a fixed acidity (pH) amount and temperature (°C). During the study the electrode maintained a constant rotation speed of 1000 rpm.

Equations of chemical reactions where the AChE is involved are shown below:

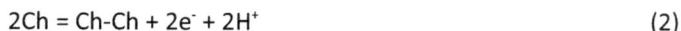

$$AChE$$

$$ACh + H_2O \Rightarrow Ch + Acetic\ Acid \tag{1}$$

$$2Ch = Ch\text{-}Ch + 2e^- + 2H^+ \tag{2}$$

During the study we measured the current produced by the oxidation reaction product Ch (Equation 2) to the enzymatic reaction (Equation 1).

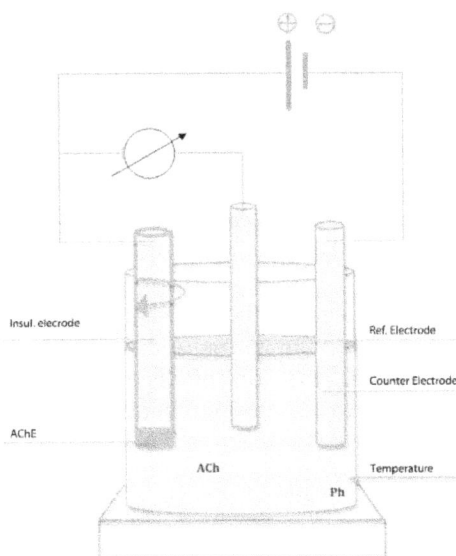

Figure 1. Acetilcholinesterase biosensor scheme

2.3. Influencing Factors

2.3.1. Substrate concentration

The behavior of the current change with the concentration of ACh is total. At higher concentrations, better results are obtained depending on the current. Current function I vs. Cs is described by a hyperbolic function:

$$I = Imax * Cs / (Cs + Km), \qquad (3)$$

where Km is a called Michaelis-Menten constant. Thus, the optimum concentration of ACh is the highest level used: 1.0 mM / L. Table 3 shows the typical experimental results.

pH	Steady-state current, µA					
	25˚C	**30˚C**	**40˚C**	**50˚C**	**60˚C**	**70˚C**
5	8.80	14.06	15.67	17.27	18.88	0.72
6	20.30	32.44	36.14	39.14	43.56	1.66
7	21.90	35.00	39.00	43.00	47.00	1.80
8	17.90	25.09	31.87	35.14	38.41	1.47
9	15.70	25.09	27.95	30.82	33.69	1.29

Table 3. Optimal results as a function of ACh concentration fixed at 1.0 mM/L
(pH 5-9 and temperature 25-70˚C)

2.3.2. pH

The response of the steady-state current, to the increase of the acidity levels, behaved in ascending form. The optimum pH greatly varies depending on temperature and Cs changes. If Cs ≤ 0.6 mM/L, and t ≤ 30˚C, the optimum pH is found at level 7; if the t > 30˚C, then it is found at level 8. If Cs > 0.6 mM / L, the optimum pH is found at level 7.The obtained data is presented in Table 4.

Substrate Concentration, mM/L	Steady-state current, µA				
	pH 5	pH 6	pH 7	pH 8	pH 9
0.2	8.56	11.22	10.65	15.79	13.13
0.4	13.32	22.27	21.32	28.36	24.55
0.6	15.68	30.98	31.98	32.35	28.62
0.8	18.11	39.85	42.63	36.87	32.39
1.0	18.88	43.56	47.00	38.41	33.69

Table 4. Optimal results in terms of pH and Cs, with a preset temperature at 60°C

2.3.3. Temperature

The response of the steady-state current to the increase in temperature is upward from 4.5 µA to 47 µA. The latter value is the maximum current found at a temperature of 60 °C. When the temperature goes from 60 to 70°C, the current decreases drastically, such that the value obtained at maximum temperature (0.382075µA, found at 70°C, pH 7, 0.2 Cs) is lower than the found at 25°C (minimum temperature) (Table 5).

Substrate Concentration, mM/L	Steady-state current, µA					
	25°C	30°C	40°C	50°C	60°C	70°C
0.2	5.60	7.63	8.68	9.60	10.65	0.38
0.4	11.20	15.26	17.37	19.20	21.31	0.76
0.6	16.30	22.89	26.06	28.80	31.97	1.14
0.8	20.00	30.52	34.75	38.40	42.63	1.52
1.0	21.90	35.00	39.00	43.00	47.00	1.80

Table 5. Optimum results in terms of temperature and Cs, with fixed pH at level 7

3. Machine Learning Algorithms

3.1. Neural Networks

Neural networks are computational models based on the structure of nerve cell connections found in the brain of living things, and likewise try to imitate their operation (Russell & Norvig, 2010; Wolfgang, 2011; Luger, 2009; Mitchell, 1997).

A neural network is composed by one or more units (neurons), which possess a series of connections serving as inputs (dendrites), outputs (axioms), or to communicate with another unit (synapses). Each connection has an associated numerical weight given. A neuron processes the input information and produces an output, which could be considered as an input to another neuron or as a final result.

A neural network operates in the following way. A vector $x = (x_1, ..., x_n)$, where $x_j \in$ R, $j = 1,...,n$, is used as input. A numerical weight w_{ji}, where i is the index of neuron, is associated with each element x_j. These values are then summed on each neuron y, and it is applied a transfer function (activation) f, which determines the output value y_i:

$$y_i = f\left(\sum_{j=1}^{n} \left(w_{ji} x_j \right) \right) \qquad (4)$$

The activation function most often used are the hyperbolic tangent function (5) or a sigmoid function (6):

$$\tanh(x) = \frac{1-e^x}{1+e^x}$$ (5)

$$sig(x) = \frac{1}{1+e^x}$$ (6)

There are different neural network architectures, developed for different types of problems. If the problem is to model (regression) or predict (classification) data, then the architectures are used as a single layer perceptron (SLP), for linearly separable problems, or multilayer perceptron (MLP), for high-dimensioned problems or not linearly separable. In addition, these architectures are based on different learning algorithms, such as the Least-Mean-Square (LMS) algorithm or the back-propagation error algorithm (BP), considered as the learning algorithm used by MLP type networks.

The MLP-BP is one of the most solicited neural network architectures in regression tasks, given its adaptability to different problems. In this case, the regression problem of approximating a possible nonlinear function $f(x)$ with a neural network $Y(x)$ BP MLP architecture, where $x \in R^n$. Figure 2 shows the scheme of a network MLP-BP.

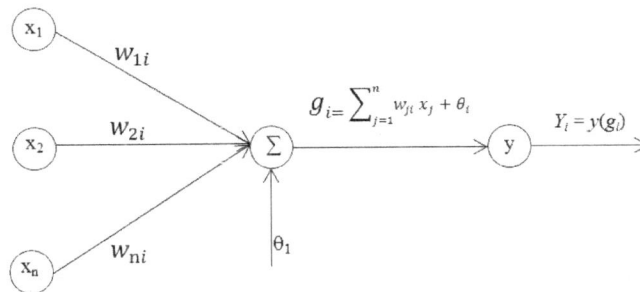

Figure 2. Example of a net MLP-BP

Inputs x_j, $j = 1... n$, of the neuron i are multiplied by the weights w_{ji} and are summed with the bias value θ_i. The result g_i is the input to the activation function Y_i. The output node i becomes:

$$Y_i = y\left(\sum_{j=1}^{n} w_{ji} x_i + \theta_i\right)$$ (7)

3.2. Support Vector Machines

The support vector machines (SVM) are supervised learning methods that generate a mapping function from a set of pre-labeled training data. The mapping can be either a classification or a regression function.

In a classification problem, a mapping is used to transform the input data, which is not linearly separable in the original space, to a space of greater dimension, which happens to be separable. The produced model depends only on a subset of the original input data whose characteristic is that it creates a boundary that separate one data class from the others. The data belonging to that subset are called *support vectors*. This boundary named margin assure us a maximum distance between the different classes (Figure 3).

The goal of the SVM is to create a computational model to predict the class label of new data samples.

In addition to their solid mathematical foundation based on statistical learning theory, the SVMs have shown a highly competitive performance in many applications such as bioinformatics, text mining, face recognition, image processing, which have been established the SVM as one of the basic tools of machine learning.

A regression problem is expressed in terms of the MSV as follows:

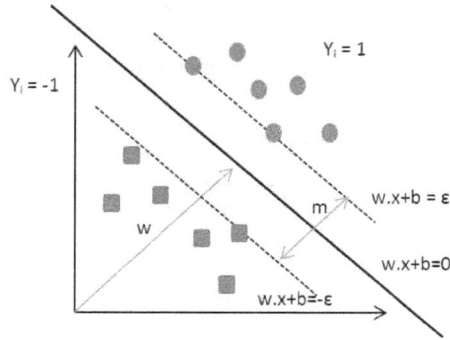

Figure 3. Classification of linearly separable data

There are given a training data set **D** = {x_i, y_i}, $i = 1... l$, of input vectors $x_i \in \mathbf{R}^n$ and their labels $y_i \in \mathbf{R}^1$. Through parameter C > 0 and ε > 0, the standard form of the SVM applied to a regression problem is (Vapnik, 1998; Chang & Lin, 2011):

$$min_{\alpha,\alpha^*} \frac{1}{2}(\alpha-\alpha^*)^T Q(\alpha-\alpha^*)+\varepsilon\sum_{i=1}^{l}(\alpha_i+\alpha_i^*)+\sum_{i=1}^{l} y_i(\alpha_i-\alpha_i^*) \tag{8}$$

s. t.

$$e^T(\alpha-\alpha^*)=0, \tag{9}$$

$$0\leq\alpha_i,\alpha_i^*\leq C, i=1,...,l, \tag{10}$$

where

$$Q_{ij}=K(x_i,x_m)\equiv\phi(x_i)^T\phi(x_m) \tag{11}$$

After solving the problem (8), the approximation function is:

$$g_{(x)}=\sum_{i=1}^{l}(-\alpha_i+\alpha_i^*)K(x_i,x)+b \tag{12}$$

where ε is a predefined constant that controls the noise tolerance. With the insensitive loss function ε, the objective is to find the function $g_{(x)}$ whose deviation is (at most) the value of the loss function ε from obtained etiquettes y_i for all training data, which should be as flat as possible.

In other words, the regression algorithm does not affect errors as long as they are less than ε, but any deviation greater than ε is not accepted.

The defined function:

$$K(x_i, x_m) \equiv \phi(x_i)^T (x_m) \tag{13}$$

is called kernel function (Figure 4). Although researchers frequently propose new kernel functions, we suggest the use of the following (Pedroza, 2007; Hammel, 2009; Hsu, Chang, & Lin, 2003; Chang & Lin, 2011):

- Linear: $K(x_i, x_m) = x_i^T x_m$.

- Polynomial: $K(x_i, x_m) = (x_i^T x_m + r)^d$, d > 0.

- Radial Basis Function: $K(x_i, x_m) = \exp(-\gamma \| x_i - x_m \|^2)$, $\gamma > 0$.

- Sigmoidal: $K(x_i, x_m) = \tanh(\gamma x_i^T x_m + r)$.

Note: *r* and *d* are specific parameters of the functions.

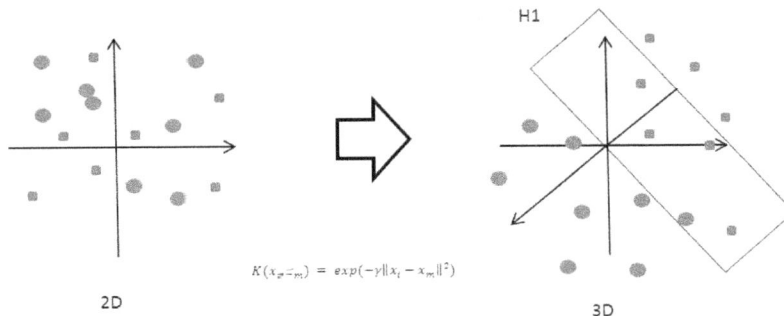

$$K(x_i, x_m) = exp(-\gamma \| x_i - x_m \|^2)$$

2D   3D

Figure 4. Transformation scheme from a set of data using a radial basis function kernel

The kernel function that reaches the objectives previously described is not known a priori; hence the procedure for its finding has not been established. The way to meet the function ϕ is based on the performance of different tests on known ϕ functions. This procedure is called *grid search*; some functions (polynomial, radial, sigmoidal, etc.), require specific parameters, which determine the separability between classes. The consumption of computational resources depends on the dimensions of the grid search (amount of features to be compared and the amount of different parameters for each function to be used), as well as the size of the data set.

Although there are different kernel functions, it is common the use of Radial Basis Function. However it can be chosen using other kernels depending on the results obtained for a particular case.

3.3. Data PreProcessing

The 150 samples measured on the electrochemical experiment were analyzed, concluding that the differences between the ranges of the parameters (Cs, pH, T) might influence the training algorithms. For this reason a min-max [0, 1] normalization process was applied. Table 6 shows the values of the data before and after standarization.

Parameters	Original Data	Normalized Data
Cs	[0.2, 0.4, 0.6, 0.8, 1.0]	[0.0, 0.25, 0.50, 0.75, 1.0]
pH	[5, 6, 7, 8, 9]	[0.0, 0.25, 0.50, 0.75, 1.0]
T	[25, 30, 40, 50, 60, 70]	[0.0, 0.1111, 0.3333, 0.5556, 0.778 ,1.0]

Table 6. Original and Normalized Data Ranges

After normalizing the data, we proceed to divide the entire set into subsets for training, validation and testing. Because the entire data set is considered small (150 samples), it is necessary to use a resampling technique, which will also permit to separate the main assembly into three subsets of data instead of 2. The first is a subset assigned for training-validation and the second is used only for testing. The resampling technique chosen is a version of the K-Fold CV, which is repeated k-times, in this way, a larger number of samples can be trained, and obtain a better generalization error. Furthermore, this technique allows the model to be evaluated in a better way because more elements are considered for validation, and reinforces the model's behavior against "new" samples.

3.4. The Regression Model

3.4.1. ANN: Training and Validation

In order to find the neural network parameters that render the Means Square Error --i.e. no. of neurons per layer, no. of internal layers, learning rate and learning algorithm, a grid search process was carried out. Considering that the search process was long, the combination of parameters for the algorithm calibration has been reduced based on previous works about approximations to functions of biological characteristics or obtained through the use of biosensors (Hsu et al, 2003).

ANN Configuration:

Layers	5
Neurons	30 per each layer
Training	Levenberg-Marquadt Algorithm
Learning	Gradient Descendent
Resampling technique	k x k-Fold CV, $k \in (5, 10)$
Performance measure	Mean Square Error (MSE)

The presented configuration offered the best results in terms of the time consumption, resources and good performance measure during training. Other prominent configurations improve the network performance on an insignificant manner; however because of their trend to consume considerable amounts resources, they were discarded. The selected configuration is programmed with 30 neurons in each of the five layers. Levenberg-Marquadt Algorithm was used to optimize the learning process. To improve the evaluation of the model a K x K-Fold CV resampling technique (Refaeilzadeh, Tang & Liu, 2009) with two different values of K (5, 10) was used.

3.4.2. SVM: Training and Validation

After the preprocessing and data division, a grid search process was applied using an exponential increasing of the parameters, as it is suggested in (Hsu et al., 2003), aiming to find the best SVR algorithm configuration in the shorter time. Since the performance of this algorithm depends on the kernel function employed, several searches were conducted, one for each type of kernel: linear, polynomial, radial basis and sigmoid. The best results were showed with a radial basis kernel function. Table 7 shows the best results of each of the 4 kernels.

The search process using the radial basis kernel was performed on 3 parameters: the error penalty (C), insensitive loss function (ε) and the radial kernel parameter (α). The resampling technique was used in this step.

	Linear	Polynomial	Radial Base Function	Sigmoid
Configuration	$c = 29$ $\varepsilon = 22.21$	$c = 29$ $\varepsilon = 22.21$ $\alpha = 0.075$ $d = 2$	$c = 512$ $\varepsilon = 4.6268$ $\alpha = 0.075$	$c = 29$ $\varepsilon = 22.21$
MSE, 5-Fold CV	40.53	26.75	5.41	7.83
MSE, 10-Fold CV	84.72	53.78	10.84	15.67

Table 7. Best SVR kernel results

The performance comparison between the algorithms mentioned above ensures that the configuration presented is the best possible for the SVR algorithm.

SVM configuration:

C	$2^9 = 512$
α	$2^{2.21} = 4.6268$
ε	0.075
Kernel	Radial Base Function
Resampling technique	$k \times k$-Fold CV, $k \in (5, 10)$
Performance measure	MSE

3.3. Neural Networks

The tests on the ANN-MLP model are realized to make a simulation of CS parameters, pH and T using the test data set and the ANN-MLP model selected during Grid Search. Figure 5 shows a comparison of the original data and the data predicted by the model ANN-MLP. A simulation of the original data and the data predicted by the ANN-MLP is exposed on Figures 6a and 6b, respectively.

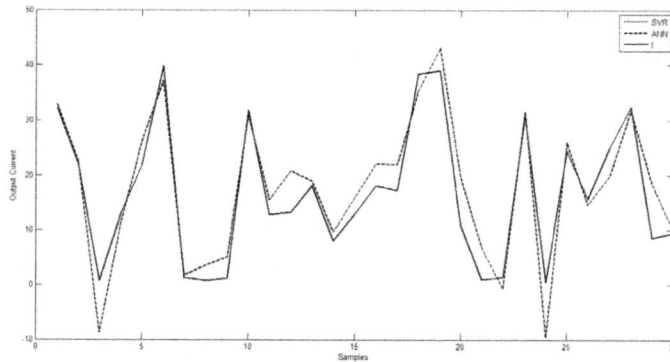

Figure 5. Graphs of the resulting current test data vs. the data predicted by the ANN-MLP model

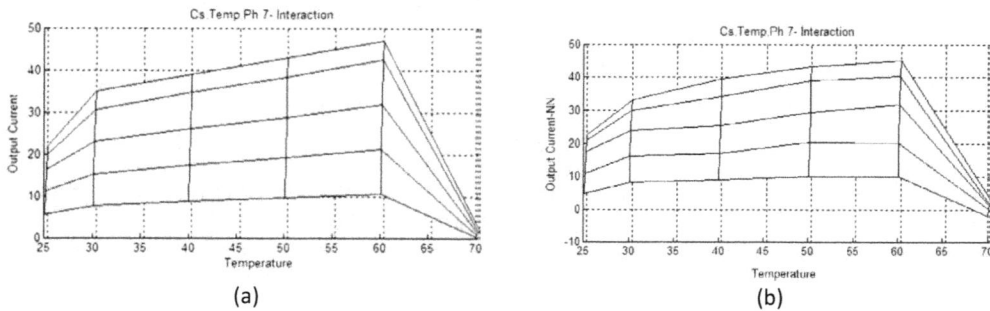

(a)

(b)

Figure 6. Comparative graphs of the ANN-MLP simulated data and the experimental data:
(a) Experimental data, (b) ANN-MLP predicted data

3.4. SVM

Similarly to the ANN-MLP, the test on the SVR model consist of the prediction of the test data set, using the best configuration found by means of a grid search process; and then, compare this predicted data set with the original test data. Table 8 shows a part of the comparison of test data and simulation using the SVR model.

No	Test Data I, µA	Simulation I, µA	Error
1	32.35	31.7865191	0.56348089
2	22.11	22.8484843	0.73848435
3	0.72	-3.47229928	4.19229928
4	13.13	14.0527819	0.9227819
5	21.9	24.0532271	2.1532271
6	39.85	40.5384729	0.68847285
7	1.32	1.76999286	0.44999286
8	0.74	3.11594287	2.37594287
9	1.16	1.50589061	0.34589061

Table 8. SVR Model best results

The comparison of SVR against the test output parameters from the original data is shown on Figure 7. A simulation of the original data and the data predicted by the SVR is exposed on Figures 8a and 8b, respectively.

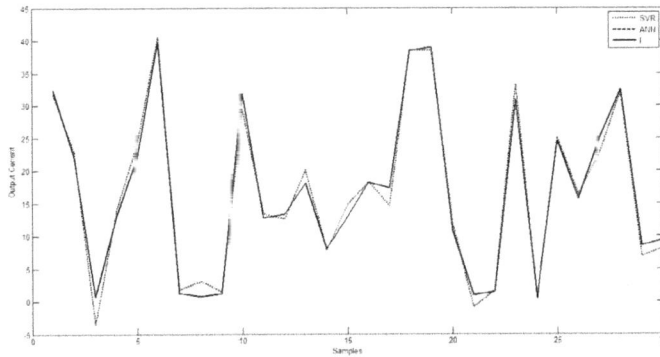

Figure 7. Current data resulting from the test set vs. the data predicted by the SVR

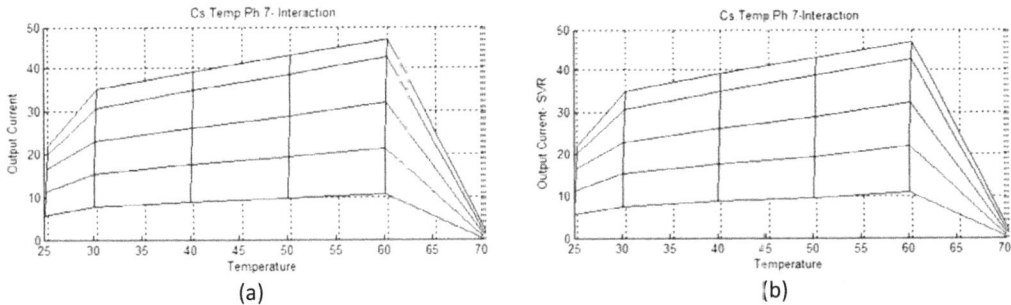

Figure 8. Comparative plots of the SVR simulated data and the experimental data:
(a) Experimental data, (b) SVR predicted data.

A comparative analysis was realized by observing the graphs of the test data modeled by the algorithms, as well as comparing errors that have the simulations of the test data and the MSE that each model has in total. Figure 9 shows the experimental test data and the data simulated by ANN-MLP and SVR models.

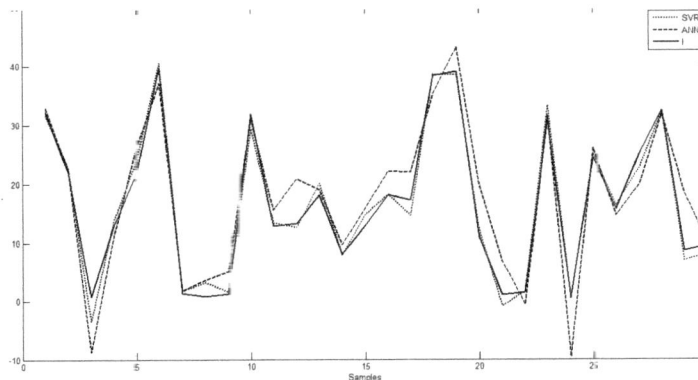

Figure 9. Comparison of experimental test data vs. data predicted by the ANN-MLP and SVR models

Table 9 shows the quantitative comparison between the test data simulated by the models and experimental data, based on the MSE of each model, 3 samples were taken randomly from the set of test data. Figure 9 and Table 9 show that both models provide good approximation function, but the SVR is the best fitting model, so it is considered the best.

No	Test Data Samples				ANN MLP		SVR	
	C_s, mmol	pH	t, °C	I, µA	I, µA	Error,	I, µA	Error,
1	0.2	6	30	8.03	9.68	1.65	7.84	0.19
2	0.8	7	50	18.15	22.21	4.06	18.26	0.09
3	0.4	7	40	38.41	35.27	3.14	38.56	0.15
Error (MSE), %					-	2.95	-	0.143

Table 9. Models quantitative comparison

In order to develop a simulation using more realistic features, a new set of samples has been generated, applying the minimum measurable level of ACh, pH and t. (Table 10).

Feature	Scale	Increment
Substrate Concentration	$0.2 \leq C_s \geq 1.0$	0.01
pH	$5 \leq pH \geq 9$	0.01
Temperature	$25 \leq t \geq 70$	0.1

Table 10. Scale simulation features

The samples generation process should consider the following:

- The data range of new samples should be the same as that of the original data.

- The generated samples should be normalized in the same way that the experimental data, using the same set of equations.

This must be hold as a way to maintain a defined structure, and avoid data extrapolation problems, causing a poor spread of new samples and therefore a higher error rate in the simulation.

Recalling that the number of samples used to develop the model were 150, with the new simulation levels, 15 467 031 samples were generated. However, in previous sections it was shown the dependency of the response variable with the Cs value; at higher concentrations, the amperage was increased and vice versa. In this way, and in order that the simulation can be analyzed qualitatively, the process was made from the perspective of T and pH, leaving the Cs value at 1.0 mM/L. For this reason, the samples were reduced to 225, 951, decreasing the samples generation, normalization and prediction time of the regression model.

To evaluate the simulation, a comparative analysis between simulated samples and experimental data were performed. Figures 10 and 11 show the minimum scale feature simulation, with Cs = 1.0 mM/ L. The results show that the function keeps the behavior of the experimental data with minor irregularities which do not compromise their performance.

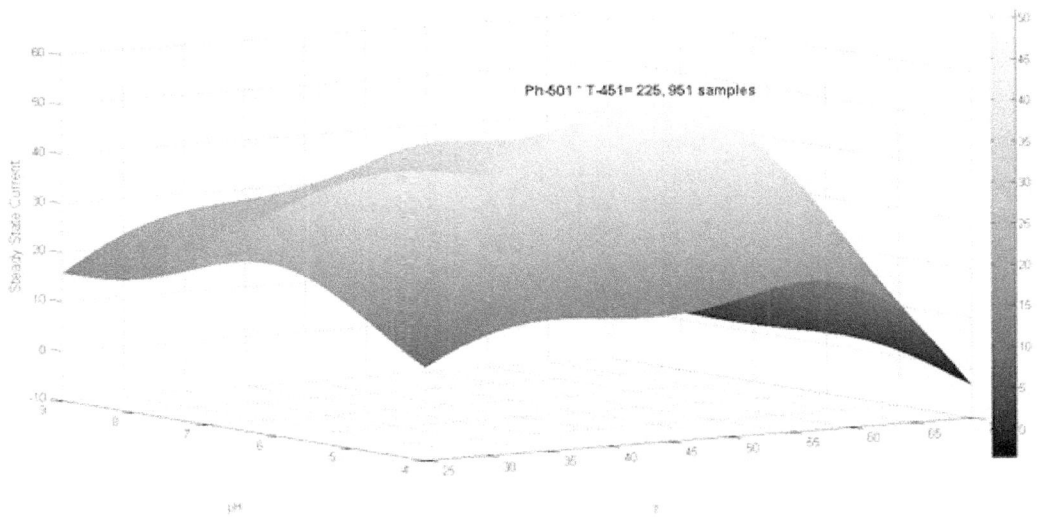

Figure 10. Minimum scale feature simulation

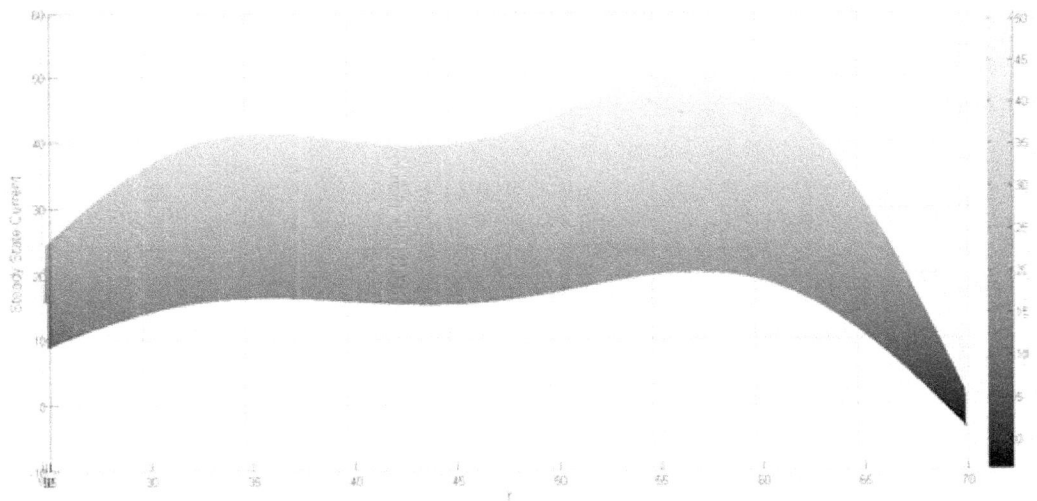

Figure 11. Comparison between original and simulated data: Temperature T perspective

In order to find the sample that generates the highest response value, a search through the whole simulation samples was made. Using this simulation, intervals for locating the maximum response value were generated; shown on Table 11.

Interval (µA)	Temperature [0.1]	pH [0.01]	# of samples in the intervals
±0.001	56.9	6.03, 6.04	2
±0.01	56.8≤ t ≤ 57.1	6.00≤ pH ≤ 6.07	24
±0.02	56.7≤ t ≤ 57.2	5.99≤ pH ≤ 6.08	48
±0.05	56.5≤ t ≤ 57.3	5.96≤ pH ≤ 6.11	117
±0.1	56.3≤ t ≤ 57.5	5.93≤ pH ≤ 6.14	232
±0.2	56.0≤ t ≤ 57.8	5.88≤ pH ≤ 6.19	472
±0.5	55.5≤ t ≤ 58.3	5.79≤ pH ≤ 6.29	1,117
±1.0	54.8≤ t ≤ 58.8	5.69≤ pH ≤ 6.41	2353
±2.0	53.8≤ t ≤ 59.6	5.55≤ pH ≤ 6.58	4810
±5.0	51.40≤ t ≤ 61.1	5.28≤ pH ≤ 6.98	12,896

Table 11. Optimum interval location

The two samples presented in the range ± 0.001 locate the maximum value of the response variable to a detectable level, and the maximum is achieved with the combination:

$$C_s = 1.0 \text{ mM/ L,} \qquad\qquad T = 56.9°C, \qquad\qquad pH = 6.03,$$

generating a steady-state current of 50.9502 µA, and the second sample that have the same configuration with the exception of pH(6.04), generates approximately the same current, with a difference of 0.000004 µA. Figure 12 highlights the area within the range ± 1.0 µA, and locates the maximum current value expressed through the simulation of samples (Garcia, Burtseva, Stoytcheva & Gonzales, 2011).

4. Conclusions

This research resolves the doubts found in the literature about learning of biological functions provided by the use of electrochemical sensors. Keeping the virtues of these, a detailed process is presented in the development of different learning models, and the procedure for the evaluation of results.

Experimental data are analyzed through different perspectives (Cs, pH, T), also their behavior and location of the possible areas of the extreme values in the resulting current, with the aim of finding the combination of parameters that maximize the sensitivity of the determination of ACh.

According to the analysis of the state of the art, the Artificial Neural Networks are the most suitable regression models to the task at hand, i.e. to predict and analyze the parameters of a biosensor. In the same way, the SVMs offer a solid and competitive performance with the possible advantage that it has less tuning parameters than a Neural Network. Although the results obtained using Neural Networks were satisfactory, the SVMs show the best performance, i.e. function approximation, both in testing and simulation modeling process.

It must be noticed that educated practices in machine learning literature instruct to computer scientists to have a proper independent test data set as a way to correctly assess the generalization capacity of any model; therefore, new test data samples must be available.

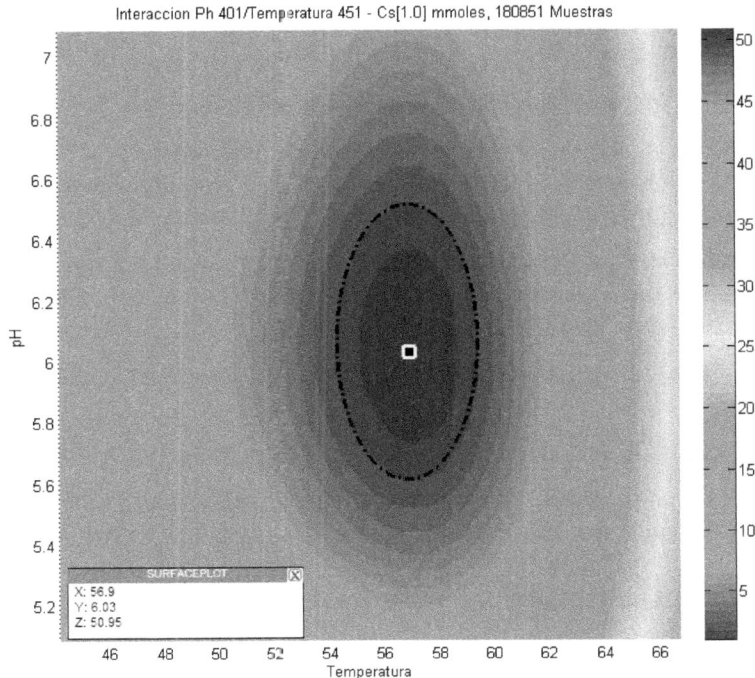

Figure 12. Location of the maximum current value

References

Alonso, G., Istamboulie, G., Ramirez-Garcia, A., Noguer, T., Marty, J., & Muñoz, R., (2010). Artificial neural network implementation in single low-cost chip for the detection of insecticides by modeling of screen-printed enzymatic sensor response. *Computers and Electronics in Agriculture, 74*, 223-229. http://dx.doi.org/10.1016/j.compag.2010.08.003

Chang, Ch.-Ch. & Lin Ch.-J., (2011). *LIBSVM: A Library for Support Vector Machines*. http://www.csie.ntu.edu.tw/~cjlin/papers/libsvm.pdf (Last access date: March 2012).

Garcia, E.R., Burtseva, L., Stoytcheva, M., & Gonzales, F.F., (2011). Predicting the behavior of the interaction of acetylthiocholine, pH and temperature of an acetylcholinesterase sensor. *LNCS-LNAI, 7094*, 583–591.

Hammel, L. (2009). *Knowledge discovery with support vector machines*. A John Wiley & Sons, Inc., Hoboken, NJ, USA, 246P.

Hsu, C.-W., Chang, C.-C., & Lin, C.-J. (2003). *A practical guide to support vector classification*. Technical report, Department of Computer Science, National Taiwan University. http://www.csie.ntu.edu.tw/~cjlin/papers/guide/guide.pdf (Last access date: March 2012).

Luger, G.F., (2009). *Artificial Intelligence: structures and strategies for complex problem solving*. Pearson Addison-Wesley, 754.

Mell, L.D., & Malloy, J.T., (1975). A Model for the Amperometric Enzyme Electrode Obtained through Digital Simulation and Applied to the Immobilized Glucose Oxidase System. *Analytical Chemistry, 47*, 299-307. http://dx.doi.org/10.1021/ac60352a006

Mitchel, T.M., (1997). *Machine Learning.* McGraw-Hill Science, 432.

Pedroza, G., (2007). Aplicación de las máquinas de soporte vectorial a reconocimiento de hablantes. Universidad Autónoma Metropolitana. México. http://cbi.izt.uam.mx/foroacademico/2007/res/cartel27.pdf (Last access date: January 2013).

Rangelova, V., Tsankova, D., & Dimcheva, N., (2010). Soft Computing Techniques in Modeling the pH and Temperature on Dopamine Biosensor. In Vermon S. Somerset (Ed.). *Intelligent and Biosensors.* Croatia: InTech. 99-122. http://dx.doi.org/10.5772/7029

Refaeilzadeh, P., Tang, L., & Liu, H., (2009). *Cross-Validation. Encyclopedia of Database Systems.* Springer, 532-538.

Russell, S.J., & Norvig, P., (2010). *Artificial Intelligence: A Modern Approach.* Prentice Hall, 1132.

Thévenot, D.R., Tóth, K., Durst, R.A., &Wilson, G.S., (1999). Electrochemical biosensors: recommended definitions and classification. *Pure and Applied Chemistry, 71(12)*, 2333-2348. http://dx.doi.org/10.1351/pac199971122333

Vapnik, V., (1998). *Statistical Learning Theory.* New York: Wiley. 736.

Wolfgang, E., (2011). *Introduction to Artificial Intelligence.* Springer. 316.

Section 3

Disturbances Modeling in Biomedical Sensors Systems

Chapter 10

Digital analysis and treatment for electromagnetic disturbances in biomedical instrumentation

Javier Enrique Gonzalez Barajas, Davis Montenegro

Universidad Santo Tomás, Colombia.

javiergonzalezb@usantotomas.edu.co, davismontenegro@usantotomas.edu.co

Doi: http://dx.doi.org/10.3926/oms.182

Referencing this chapter

Gonzalez Barajas, J.E., Montenegro, D. (2014). Digital analysis and treatment for electromagnetic disturbances in biomedical instrumentation. In M. Stoytcheva & J.F. Osma (Eds.). *Biosensors: Recent Advances and Mathematical Challenges*. Barcelona, España: OmniaScience, pp. 205-217.

1. What is a measurement's network?

The control and monitoring systems to ensure that a process can be performed successfully have become more complex in the last decades, due the inclusion of nonlinear variables; their complexity tends to be increased. All those systems have a highly dependency of feedback signals to increase his reliability, which are acquired by using different types of sensors distributed within the system to simplify the system complexity (Kopetz, 2011), transforming it into information for the control/monitoring system and to deliver useful information about the system status to the user (Kezunovic, Vittal, Meliopoulos & Mount, 2012).

The distributed measurements built a network of measurements (MeN) (Song, Qiu & Zhang, 2011; Pin-Hsuan, 2011), which consist in a group of measurements distributed within the observed system, to describe the possible functional status associated to it against the time. This kind of MeN can be found in biomedical applications (Khalili & Moradi, 2009; Young, 2009), power system analysis applications (Baran & McDermott, 2009), development of high performance communication systems (De Grande, Boukerche & Ramadan (2011), among others.

For example, to measure electrocardiographic signals it is necessary to take measurements at different parts of the patient's body; this way it will be possible, based in the combination of the acquired signals, infer which is the health status of the patient and if he needs some kind of special care due to the results of the study. The implementation of the mentioned example is shown in Figure 1.

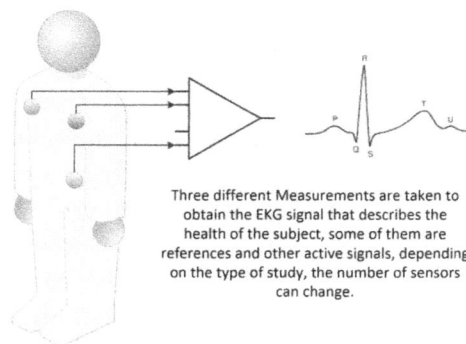

Three different Measurements are taken to obtain the EKG signal that describes the health of the subject, some of them are references and other active signals, depending on the type of study, the number of sensors can change.

Figure 1. Example of MeN in a biological system

Each measurement within the MeN has to be characterized and correlated with the other also characterized measurements; this way, the set of measurements built a base to determinate the functional states of the observed system.

Additionally to the desired measurements, there are other non-desired components that must be removed or at least considered to handle them and improve the instrument behavior. This chapter presents the generalized techniques to characterize the instrument individually, then how the MeN are built for linear and nonlinear systems. As third part, the digital signal processing (DSP) techniques for removing undesired disturbances is presented, making emphasis in biological signals, which are highly sensitive to different types of disturbances due to his magnitude and wideband.

1.1. Modeling a single instrument

A single instrument can be described as the combination of a data carrier signal with errors signals, the error signals presents different characteristics and can be modeled taking into account the static and dynamic characteristics of the measurement system (Bentley, 2005).

The general diagram to describe the structure of an instrument is shown in the Figure 2. In this structure is possible to visualize that an instrument is the result of different components and each component incorporates dynamic and static error components.

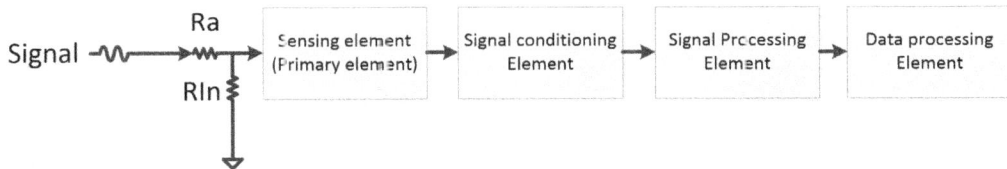

Figure 2. General structure of an instrument

It is also possible to observe that there are digital and analog components within the instrument model, the first one and more important analog component is the input impedance of the instrument (Rin), which depending on the type of signal (voltage or current), must have a value adequate to interact with the source of the signal; this is, for voltage signals the value of Rin must tend to a high impedance value, while for current signals his value must tend to zero. The losses associated to this first stage can be calculated with the following expression:

$$Loss(\%)=\left(1-\frac{Rin}{Rin+Ra}\right)*100 \tag{1}$$

The Equation 1 presents an expression to calculate the losses due the input impedance of the measurement system and the output impedance, generated by the direct or indirect contact between the primary element (sensor) and the measurement point, which can be separated by a dielectric component like air, the skin of a patient, etc.

But when the magnitude oscillates, this is changes of his magnitude against the time; some additional consideration has to be taken into account like the parasite capacitances present at the instrument input. This capacitances works like an analog filter for specific frequencies sending his spectral components to zero, changing the original spectrum of the measured signal and sometimes, reducing the amount of representative information in the acquired data. The complete model is shown in Figure 3.

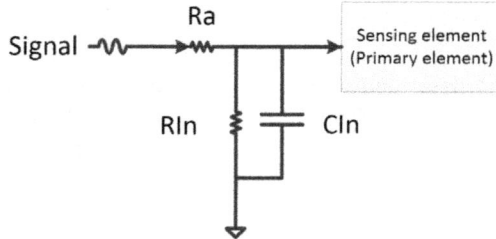

Figure 3. Complete model for the instrument impedance

With the proposed configuration the new equation to describe the losses due the input impedance of the circuit are described by the following expression:

$$Loss\,(\%)=1-\left|\frac{\dfrac{Rin}{j\omega C}}{\dfrac{Rin}{j\omega C}+\left(Rin+\dfrac{1}{j\omega C}\right)Ra}\right|*100 \tag{2}$$

Where ωC is the term to determinate which components of the input spectrum are going to be affected by the value of C, a characteristic that can be treated by using another parallel capacitances (ω = 2πf) or adding other configurations with LC or RL components.

Inside the instrument, the different stages described in Figure 2 has associated two types of errors: the systematic error and the random error (Dunn, 2005; Webster & Eren, 2013); the first one can be treated because it can be modeled and minimized using mathematical methodologies, but the second one, because of his random behavior, can only be associated to a probability distribution function where the standard deviation is defined by the manufacturing process and materials, and can be identified by the class index of the element.

This way, each part of the instrument can be described as follows:

$$Signal\,(t)=f(t)+Es+Ea\,(t) \tag{3}$$

Where f(t) is the acquired signal after input impedance, Es is the systematic error, which many times is associated to the magnitude of the measurement and Ea(t), is the random error. To calculate the total error associated to the measurement system due the presence of the random error, and the low probability that all the elements of the instruments are going to fail at the same magnitude at the same time instant, the following expression can be used:

$$Total\ Error = \sqrt{E_1^2 + E_2^2 + E_3^2 ... + E_n^2} \tag{4}$$

Where E_n are the errors (class index) associated to each component of the instrument. This is because the systematic error can be corrected by the user after a calibration procedure in site; this way the total error can be included in the range of the instrument like an expected variation in the measurement (Ej: 0-85°C ± 1%).

1.2. Modeling a measurement's network

A MeN is a set of measurements to describe the system functional state, there are many applications on MeN oriented to describe linear and nonlinear systems. This relationship between the different measurements is made using matrix arrangements, which involves the taken measurements to calculate other variables that can be estimated in the same or other units. One example of this is the hydrostatic tank gauging (HTG) used to determinate the level of a liquid within a closed recipient based in pressure and temperature measurements. The general set of measurements to do this build this model is the following:

$$\begin{bmatrix} 1/h12 & -1/h12 & 0 & 0 \\ 1/Den(z-1) & 0 & -1/Den(Z-1) & 1 \\ 1/At & 0 & -1/At & 0 \end{bmatrix} \begin{bmatrix} P1 \\ P2 \\ Ps \\ hi \end{bmatrix} + e = \begin{bmatrix} Den(z) \\ Level \\ Mass \end{bmatrix} \tag{5}$$

Where *Den(z)* is the density of the liquid at the moment of measurement, *Den(Z-1)* is the density of the liquid at the previous instant of measurement, *At* is the transversal area of the tank, *h12* is the high between the pressure sensors *p1* and *p2*, *e* is the error vector and *hi*, is the high between the bottom of the tank and the sensor *p1*. In this model it is possible to add more measurements like temperature to improve the accuracy of the results.

2. Disturbances modeling

2.1. Disturbances classification

Electrical equipment adds disturbances that affect directly the MeN applied to biomedical instrumentation system; the electrical disturbances are the most common found due the fact that electric power sources are necessary to instrumentation devices work (Chatterjee & Miller, 2010).

2.2. Mathematical models to describe disturbances

If x(t) is a signal acquired by a MeN System, it is composed by a signal under study s(t) and the disturbance signal r(t) (Equation 6).

$$x(t) = s(t) + r(t) \tag{6}$$

In this case, r(t) is a disturbance generated by a power electric system. The Equation 7 shows a basic mathematical model to describe it where A is the signal amplitude and F is his frequency, (60 HZ for american system and 50 Hz for european system) (Kim, Ku, Kim, Kim & Nam (2007).

$$r(t) = A * \cos \ (2 * \pi * F * t) \tag{7}$$

The power system signal can compose by more than one frequency component, a reason for why it is necessary his caracterization in the frequency domain. Figure 4 shows the spectrum of two types of disturbances: constant frequency and variable frequency

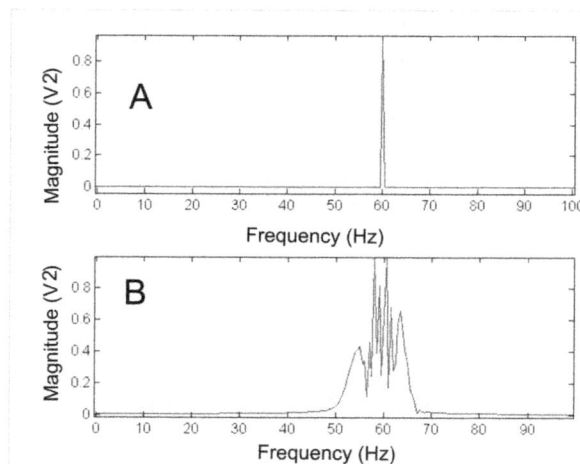

Figure 4. Electric power perturbation: constant frequency (A) and variable frequency (B)

2.3. Detection of disturbances

Through the Fourier Fast Transform (FFT) it is possible to study the acdition process between a biomedical signal and the disturbance with variable frequency. In this case, the biomedical signal is a electrocardiographic one. Figure 5 shows the simulation.

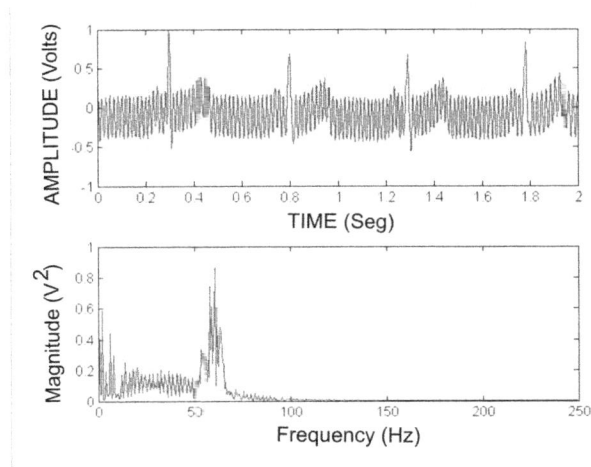

Figure 5. Simulation of addition process between a electrocardicgraphic signal and a power electrical disturbance

3. Design and implementation of disturbances simulation

The power electrical system has alterations caused by the non-adequate operation of the electrical installations. These alterations must be implemented in the simulation of power electrical disturbances.

3.1. Continuous or discrete time?

The electrical disturbance is a time continuous signal, but it can be simulated using a time discrete signal. It is important to include in the model characteristics like voltage drops, voltage cuts, over voltage and the inclusion of steady state and transient frequency components in the discrete signal.

3.2. Alternatives for implement a simulation

An electrical power perturbation with harmonics components can be simulated using the equation in 8.

$$r(t) = A_0 * \cos(2*\pi*60*t) + A_1 * \cos(2*\pi*180*t) + A_2 * \cos(2*\pi*300*t) \qquad (8)$$

The Figure 6 shows the simulation of a biomedical signal distorted by an electrical power signal with harmonic components.

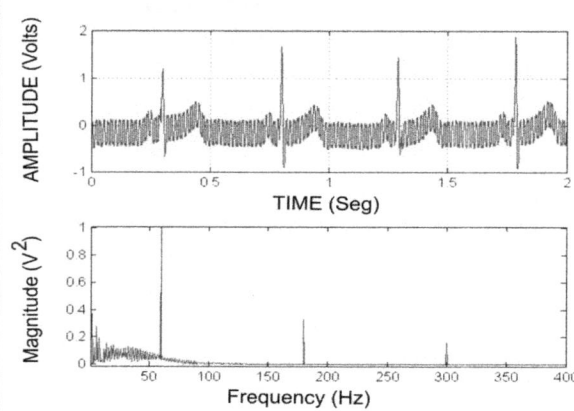

Figure 6. Electrocardiographic signal distorted by an electrical power disturbance based in harmonics components

3.3. The system dynamics and the code to simulate

The simulation of an electric power perturbation using a sampling frequency 500 Hz and 500 samples, can be implemented using Matlab and the following program lines:

```
fs=500;
Ts=1/fs;
N=500;
n=1:N;
t=(n-1)/fs;
r=0.5*sin(2*pi*60*n*Ts);
```

An electric power perturbation with sinusoidal variation of its frequency value:

```
R=0.7*sin(2*pi*0.8*n*Ts)+60;
for n=1:500
r(n)=0.4*sin(2*pi*R(n)*n*Ts);
end
```

An electric power perturbation with harmonics components:

```
fs=1500;
Ts=1/fs;
N=3000;
n=1:N;
t=(n-1)/fs;
r=0.3*sin(2*pi*60*n*Ts) +0.1*sin(2*pi*180*n*Ts)+0.05*sin(2*pi*300*n*Ts);
```

4. Classic methodologies for removing disturbances in measurement networks

A classical solution is a notch filter using the Equation 9. Where *a* depends of the disturbance frequency (60Hz) and the sampling frecuency (Fs) (Equation 10).

$$y(n)=x(n) + a^*x(n-1) + x(n-2) \qquad (9)$$

$$a=-2*\cos\left(\frac{60}{Fs}\right) \qquad (10)$$

The notch filter is an effective solution only if the perturbation is composed by a sinusoidal signal with 60 Hz. Figure 7 shows the performance of the notch filter applied to a disturbance with harmonics components.

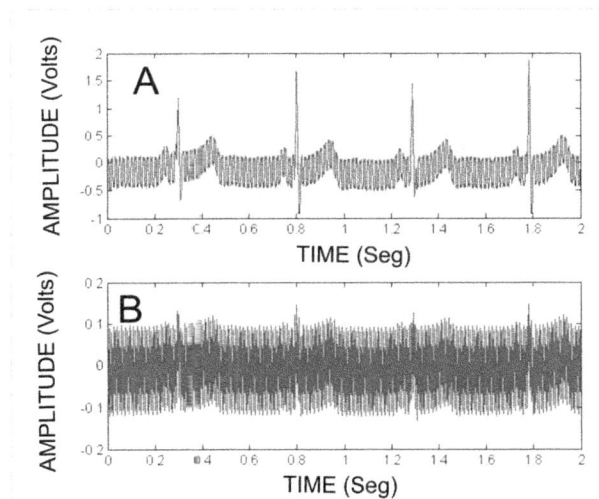

Figure 7. Electrocardiographic signal distorted by an electrical power disturbance with harmonics components (A) and the signal filtered using a notch filter (B)

5. Design and implementation of complete algorithms

An adaptive filter FIR (Figure 8) is composed of an input signal x(n) and output signal y(n). The signal e(n) is calculated using the difference between y(n) and the desired signal d(n).The adaptation ruler uses the coefficients Wk applied to the cancellation of the error signal e(n).

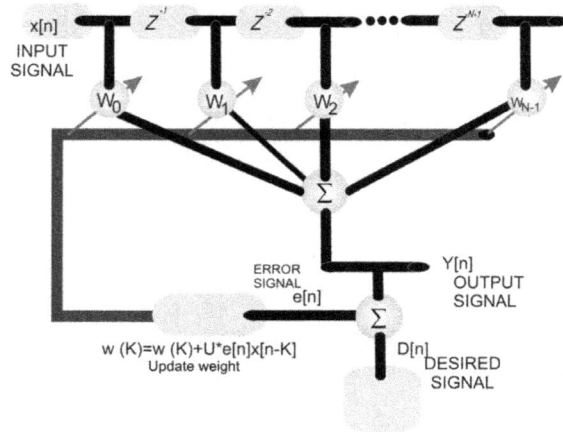

Figure 8. Block diagram of adaptative FIR filter

5.1. The inclusion of digital filters in simulated systems

In Matlab is possible the adaptive FIR filter implementation using the following program lines

```
B=zeros(1,50);
B(1)=1;
L=length(B);
% Cálculo del paso de coeficiente de μ
StepU=1/(2*sum(x.^2));

for n=L+1:N
c=0;
for k=1:L
    c=c+B(k)*x(n-k);
end
y(n)=c;
e(n)=d(n)-y(n);

for k=1:L
B(k)=B(k)+e(n)*StepU*x(n-k);
end

end
```

The coefficients of the adaptive FIR filter can produce a transfer function applied to the electric disturbance attenuation. Figure 9 shows the frequency response of the transfer function applied to disturbance with harmonics components.

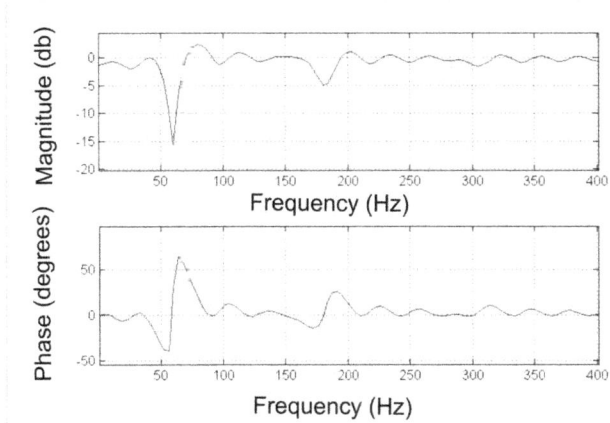

Figure 9. Frequency response of the adaptive FIR filter applied to electric power disturbance with harmonics components

Figure 10 shows the electrocardiographic signal distorted using an electrical power disturbance with harmonics components filtered by an adaptive FIR filter.

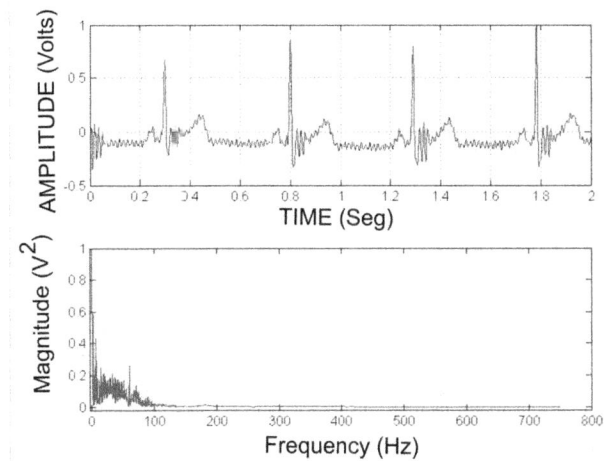

Figure 10. Electrocardiographic signal filtered by an adaptive FIR filter

5.2. The selection of parameters for digital treatment of the disturbances

To implement a system for the treatment of disturbances, it is necessary to have criteria for selecting the parameters of the algorithms.

In the implementation of an adaptive FIR filter is important the selection of the parameter μ. It is the adaptation coefficient (Equation 11).

$$\mu = \frac{1}{2 * \sum_{n=0}^{N-1} |x(n)|^2}$$

(11)

6. Conclusions

It has been presented a case of application where the different signal components within a biomedical instrumentation are digitally treated, due the presence of different types of errors and sources, it is necessary to use multiple methods to minimize them.

The MeN represents a complex system that allows improving the performance of an instrumentation system, but it is necessary to consider that multiple errors and disturbances are involved in this complex system to achieve the expected results.

The biomedical instrumentation needs sensor networks applied to electrophysiological signal acquisition. The electrophysiological signal can be distorted by disturbances originated in the electrical system. This chapter tried models and methodologies to simulate the distortions and the strategies to mitigate the influence in biomedical sensors systems.

References

Baran, M., & McDermott, T.E. (2009). Distribution System State Estimation Using AMI Data. *IEEE/PES Power Systems Conference and Exposition*, Sea tle, WA, 1-3. http://dx.doi.org/10.1109/PSCE.2009.4840257

Bentley, J.P. (2005). *Principles of measurement systems.* Pearson Prentice Hall.

Chatterjee, S., & Miller, A. (2010). *Biomedical Instrumentation Systems.* Delmar Cengage Learning.

De Grande, R.E., Boukerche, A., & Ramadan, H.M.S. (2011). Decreasing Communication Latency through Dynamic Measurement, Analysis, and Partitioning for Distributed Virtual Simulations. *IEEE Transac tions on Instrumenta tion and Measurement, 60,* 81-92. http://dx.doi.org/10.1109/TIM.2010.2065730

Dunn, W. (2005). *Fundamentals of Industrial Instrumentation and Process Control.* McGraw-Hill Education.

Kezunovic, M., Vittal, V., Meliopoulos, S., & Mount, T. (2012). The Big Picture: Smart Research for Large-Scale Integrated Smart Grid Solutions. *IEEE Power and Energy Magazine, 10,* 22-34. http://dx.doi.org/10.1109/MPE.2012.2196335

Khalili, Z., & Moradi, M.H. (2009). Emotion recognition system using brain and peripheral signals: Using correlation dimension to improve the results of EEG. *International Joint Conference on Neural Networks (IJCNN)*, 1571-1575.

Kim, K.J., Ku, J.H., Kim, I.Y., Kim, S.I., & Nam, S.W. (2007). Notch filter design using theα-scaled sampling kernel and its application to power line noise removal from ECG signals. *International Conference on Control, Automation and Systems (ICCAS)*, 2415-2418.

Kopetz, H. (2011). *Real-Time Systems.* Second ed. New York, NY 10013, USA: Springer Science+Business Media.

Pin-Hsuan, C. (2011). Smart browser: Network measurement system based on perfSONAR framework. *Network Operations and Management Symposium (APNOMS), Asia-Pacific*, 1-4.

Song, H., Qiu, L., & Zhang, Y. (2011). NetQuest: A Flexible Framework for Large-Scale Network Measurement. *IEEE/ACM Transactions on Networking, 17,* 106-119. http://dx.doi.org/10.1109/TNET.2008.925535

Webster, J.G., & Eren, H. (2013). *Measurement, Instrumentation, and Sensors Handbook. Second Edition: Spatial, Time, and Mechanical Variables.* Taylor & Francis Group.

Young, D.J. (2009). Development of wireless batteryless implantable blood pressure-EKG-core body temperature sensing microsystem for genetically engineered mice real time monitoring. *2009 IEEE International Conference on Nano/Molecular Medicine and Engineering (NANOMED)*, 259-264. http://dx.doi.org/10.1109/NANOMED.2009.5559075

About the editors

MARGARITA STILIANOVA STOYTCHEVA

Autonomous University of California, Mexico.

margarita.stoytcheva@uabc.edu.mx

Margarita Stoytcheva graduated from the University of Chemical Technology and Metallurgy of Sofia, Bulgaria, with titles of Chemical Engineer and Master of Electrochemical Technologies. She has a Ph.D. and DSc. degrees in chemistry and in technical sciences. She has acted in research and teaching in several Universities in Bulgaria, Algeria, and France. From 2006 to the present she has participated in activities of scientific research, technological development, and teaching at Mexico, at the University of Baja California, Institute of Engineering, Mexicali, as a full time researcher. Since 2008 she has been a member of the National System of Researchers of Mexico. Her interests and areas of research are analytical chemistry, electrochemistry, and biotechnology.

JOHANN F. OSMA

University of Los Andes, Colombia.

jf.osma43@uniandes.edu.co

Johann F. Osma is an electronic Engineer with MSc. in microelectronics from the University of los Andes (Colombia). He started his research career at the Microelectronics Research Center – CMUA- of the same university in 2001 under the field of micro and nanotechnologies. Later, he continued his academic formation with a MSc. and PhD in Chemical and Environmental Engineer at Universitat Rovira i Virgili (Spain) in the field of bio-nanotechnology. Since 2010 he joined the University of los Andes as faculty staff, where he started the research field on biosensors and microfluidic systems and manages a clean-room specialized on those topics. He has published multiple articles and book chapters, a book, and works as a frequent reviewer and member of editorial boards of some journals.